AF327419

Selections from

Statistics for Business and Economics

Eighth Edition

University of Pretoria

Anderson, Sweeney, Williams

COPYRIGHT © 2002 by South-Western College Publishing, a division of Thomson Learning Inc. Thomson Learning™ is a trademark used herein under license.

Printed in the United States of America

South-Western College Publishing
5194 Natorp Blvd
Mason, Ohio 45040
USA

For information about our products, contact us:
Thomson Learning Academic Resource Center
1-800-423-0563
http://www.swcollege.com

International Headquarters
Thomson Learning
International Division
290 Harbor Drive, 2nd Floor
Stamford, CT 06902-7477
USA

UK/Europe/Middle East/South Africa
Thomson Learning
Berkshire House
168-173 High Holborn
London WCIV 7AA

Asia
Thomson Learning
60 Albert Street, #15-01
Albert Complex
Singapore 189969

Canda
Nelson Thomson Learning
1120 Birchmount Road
Toronto, Ontario MIK 5G4
Canada
United Kingdom

ALL RIGHTS RESERVED. No part of this work covered by the copyright hereon may be reproduced or used in any form or by any means—graphic, electronic, or mechanical, including photocopying, recording, taping, Web distribution, or information storage and retrieval systems—without the written permission of the publisher.

ISBN 978-0-324-6467-8
(0-324-26467-4)

The Adaptable Courseware Program consists of products and additions to existing South-Western College Publishing products that are produced from camera-ready copy. Peer review, class testing, and accuracy are primarily the responsibility of the author(s).

INDEX NUMBERS

CONTENTS

STATISTICS IN PRACTICE

U.S. DEPARTMENT OF LABOR
BUREAU OF LABOR STATISTICS
Washington, D.C.

The U.S. Department of Labor, through its Bureau of Labor Statistics, compiles and distributes indexes and statistics that are indicators of business and economic activity in the United States. For instance, the Bureau compiles and publishes the Consumer Price Index, the Producer Price Index, and statistics on average hours and earnings of various groups of workers. Perhaps the most widely quoted index produced by the Bureau of Labor Statistics is the Consumer Price Index. It is often used as a measure of inflation.

In September 2000, the Bureau of Labor Statistics reported that the Consumer Price Index (CPI) decreased by .1% from the July level. This decrease was the first in 14 years. The Bureau noted, however, that the core rate of inflation was up .2% in August. The core rate excludes the volatile food and energy components of the CPI and is sometimes regarded as a better indicator of inflationary pressures. The energy index fell by 2.9% and the food index increased by .2%.

Many economists and analysts proclaimed that the CPI report confimed that the United States was in a noninflationary era. In addition to the CPI report, they pointed to very little upward pressure on wages. The Labor Department had also reported that average weekly earnings adjusted for inflation only rose .1% in August after being unchanged in each of the prior two months.

Another indicator that inflationary pressures would be low for the near future was given by the

The Bureau of Labor Statistics computes the Consumer Price Index from data collected on the sale of goods and services. © CORBIS.

Producer Price Index. It measures price changes in wholesale markets and is often seen as a leading indicator of changes in the Consumer Price Index. The Producer Price Index fell by .2% in August. Federal Reserve policymakers had raised the overnight bank lending rate six times since June 1999 but, with this data that prices were not rising rapidly, were expected to leave interest rates unchanged for the near future.

In this chapter we will see how various indexes, such as the Consumer and Producer Price Indexes, are computed and how they should be interpreted.

Each month the U.S. government publishes a variety of indexes that are designed to help individuals understand current business and economic conditions. Perhaps the most widely known and cited of these indexes is the Consumer Price Index (CPI). As its name implies, the CPI is an indicator of what is happening to prices consumers are paying for items purchased. Specifically, the CPI measures changes in price over a period of time. With a given starting point or *base period* and its associated index of 100, the CPI can be used to compare current period consumer prices with those in the base period. For example, a CPI of 125 reflects the condition that consumer prices as a whole are running approximately 25% above the base period prices for the same items. Although relatively few individuals know exactly what this number means, they do know enough about the CPI to understand that an increase means higher prices.

Even though the CPI is perhaps the best-known index, many other governmental and private-sector indexes are available to help us measure and understand how economic conditions in one period compare with economic conditions in other periods. The purpose of this chapter is to describe the most widely used types of indexes. We will begin by constructing some simple index numbers to gain a better understanding of how indexes are computed.

17.1   PRICE RELATIVES

TABLE 17.1

UNLEADED GASOLINE COST

Year	Price per Gallon ($)
1984	1.21
1985	1.20
1986	.93
1987	.95
1988	.95
1989	1.02
1990	1.16
1991	1.14
1992	1.13
1993	1.11
1994	1.11
1995	1.15
1996	1.23
1997	1.23
1998	1.06

Source: U.S. Energy Administration, *Monthly Energy Review.*

The simplest form of a price index shows how the current price per unit for a given item compares to a base period price per unit for the same item. For example, Table 17.1 reports the cost of one gallon of unleaded gasoline for the years 1984 through 1998. To facilitate comparisons with other years, the actual cost-per-gallon figure can be converted to a **price relative,** which expresses the unit price in each period as a percentage of the unit price in a base period.

$$\text{Price relative in period } t = \frac{\text{Price in period } t}{\text{Base period price}} (100) \qquad \textbf{(17.1)}$$

For the gasoline prices in Table 17.1 and with 1984 as the base year, the price relatives for one gallon of unleaded gasoline in the years 1984 through 1998 can be calculated. These price relatives are listed in Table 17.2. Note how easily the price in any one year can be compared with the price in the base year by knowing the price relative. For example, the price relative of 77 in 1986 shows that the gasoline cost in 1986 was 23% below the 1984 base-year cost. Similarly, the 1998 price relative of 88 shows a 12% decrease in gasoline cost in 1998 from the 1984 base-year cost. Price relatives, such as the ones for unleaded gasoline, are extremely helpful in terms of understanding and interpreting changing economic and business conditions over time.

17.2   AGGREGATE PRICE INDEXES

TABLE 17.2

PRICE RELATIVES FOR ONE GALLON OF UNLEADED GASOLINE (1984–1998)

Year	Price Relative (Base 1984)
1984	(1.21/1.21)100 = 100.0
1985	(1.20/1.21)100 = 99.2
1986	(.93/1.21)100 = 76.9
1987	(.95/1.21)100 = 78.5
1988	(.95/1.21)100 = 78.5
1989	(1.02/1.21)100 = 84.3
1990	(1.16/1.21)100 = 95.9
1991	(1.14/1.21)100 = 94.2
1992	(1.13/1.21)100 = 93.4
1993	(1.11/1.21)100 = 91.7
1994	(1.11/1.21)100 = 91.7
1995	(1.15/1.21)100 = 95.0
1996	(1.23/1.21)100 = 101.7
1997	(1.23/1.21)100 = 101.7
1998	(1.06/1.21)100 = 87.6

Although price relatives can be used to identify price changes over time for individual items, we are often more interested in the general price change for a group of items taken as a whole. For example, if we want an index that measures the change in the overall cost of living over time, we will want the index to be based on the price changes for a variety of items, including food, housing, clothing, transportation, medical care, and so on. An **aggregate price index** is developed for the specific purpose of measuring the combined change of a group of items.

Consider the development of an aggregate price index for a group of items categorized as normal automotive operating expenses. For illustration, we limit the items included in the group to gasoline, oil, tire, and insurance expenses.

Table 17.3 gives the data for the four components of our automotive operating expense index for the years 1984 and 1998. With 1984 as the base period, an aggregate price index for the four components will give us a measure of the change in normal automotive operating expenses over the 1984–1998 period.

An unweighted aggregate index can be developed by simply summing the unit prices in the year of interest (e.g., 1998) and dividing that sum by the sum of the unit prices in the base year (1984). Let

$$P_{it} = \text{unit price for item } i \text{ in period } t$$
$$P_{i0} = \text{unit price for item } i \text{ in the base period}$$

TABLE 17.3   DATA FOR AUTOMOTIVE OPERATING EXPENSE INDEX

| | Unit Price ($) | |
Item	1984	1998
Gallon of gasoline	1.21	1.06
Quart of oil	1.50	2.20
Tires	80.00	145.00
Insurance policy	300.00	700.00

An unweighted aggregate price index in period t, denoted by I_t, is given by

$$I_t = \frac{\Sigma P_{it}}{\Sigma P_{i0}} (100) \tag{17.2}$$

where the sums are over all items in the group.

An unweighted aggregate index for normal automotive operating expenses in 1998 ($t = 1998$) is given by

$$I_{1998} = \frac{1.06 + 2.20 + 145.00 + 700.00}{1.21 + 1.50 + 80.00 + 300.00} (100)$$

$$= \frac{848.26}{382.71} (100) = 222$$

If quantity of usage is the same for each item, an unweighted index gives the same value as a weighted index. In practice, however, quantities of usage are rarely the same.

From the unweighted aggregate price index, we might conclude that the price of normal automotive operating expenses increased 122% over the period from 1984 to 1998. But note that the unweighted aggregate approach to establishing a composite price index for automotive expenses is heavily influenced by the items with large per-unit prices. Consequently, items with relatively low unit prices such as gasoline and oil are dominated by the high unit-price items such as tires and insurance. The unweighted aggregate index for automotive operating expenses is too heavily influenced by price changes in tires and insurance.

Because of the sensitivity of an unweighted index to one or more high-priced items, this form of aggregate index is not widely used. A weighted aggregate price index provides a better comparison when usage quantities differ.

The philosophy behind the **weighted aggregate price index** is that each item in the group should be weighted according to its importance. In most cases, the quantity of usage is the best measure of importance. Hence, one must obtain a measure of the quantity of usage for the various items in the group. Table 17.4 gives annual usage information for each item of automotive operating expense based on the typical operation of a midsize automobile for approximately 15,000 miles per year. The quantity weights listed show the expected annual usage for this type of driving situation.

Let $Q_i =$ quantity of usage for item i. The weighted aggregate price index in period t is given by

TABLE 17.4

ANNUAL USAGE INFORMATION FOR AUTOMOTIVE OPERATING EXPENSE INDEX

Item	Quantity Weights*
Gallons of gasoline	1000
Quarts of oil	15
Tires	2
Insurance policy	1

*Based on 15,000 miles per year. Tire usage is based on a 30,000-mile tire life.

$$I_t = \frac{\Sigma P_{it} Q_i}{\Sigma P_{i0} Q_i} (100) \tag{17.3}$$

where the sums are over all items in the group. Applied to our automotive operating expenses, the weighted aggregate price index is based on dividing total operating costs in 1998 by total operating costs in 1984.

Let $t = 1998$, and use the quantity weights in Table 17.4. We obtain the following weighted aggregate price index for automotive operating expenses in 1998.

$$I_{1998} = \frac{1.06(1000) + 2.20(15) + 145.00(2) + 700.00(1)}{1.21(1000) + 1.50(15) + 80.00(2) + 300.00(1)}(100)$$

$$= \frac{2083}{1692.5}(100) = 123$$

From this weighted aggregate price index, we would conclude that the price of automotive operating expenses has increased 23% over the period from 1984 through 1998.

Clearly, compared with the unweighted aggregate index, the weighted index provides a more accurate indication of the price change for automotive operating expenses over the 1984–1998 period. Taking the quantity of usage of gasoline into account helps to offset the large increase in insurance costs. The weighted index shows a more moderate increase in automotive operating expenses than the unweighted index. In general, the weighted aggregate index with quantities of usage as weights is the preferred method for establishing a price index for a group of items.

In the weighted aggregate price index formula (17.3), note that the quantity term Q_i does not have a second subscript to indicate the time period. The reason is that the quantities Q_i are considered fixed and do not vary with time as the prices do. The fixed weights or quantities are specified by the designer of the index at levels believed to be representative of typical usage. Once established, they are held constant or fixed for all periods of time the index is in use. Indexes for years other than 1998 require the gathering of new price data P_{it}, but the weighting quantities Q_i remain the same.

In a special case of the fixed-weight aggregate index, the quantities are determined from base-year usages. In this case we write $Q_i = Q_{i0}$, with the zero subscript indicating base-year quantity weights; (17.3) becomes

$$I_t = \frac{\Sigma P_{it}Q_{i0}}{\Sigma P_{i0}Q_{i0}}(100) \qquad\qquad \textbf{(17.4)}$$

Whenever the fixed quantity weights are determined from base-year usage, the weighted aggregate index is given the name **Laspeyres index.**

Another option for determining quantity weights is to revise the quantities each period. A quantity Q_{it} is determined for each year that the index is computed. The weighted aggregate index in period t with these quantity weights is given by

$$I_t = \frac{\Sigma P_{it}Q_{it}}{\Sigma P_{i0}Q_{it}}(100) \qquad\qquad \textbf{(17.5)}$$

Note that the same quantity weights are used for the base period (period 0) and for period t. However, the weights are based on usage in period t, not the base period. This weighted aggregate index is known as the **Paasche index.** It has the advantage of being based on current usage patterns. However, this method of computing a weighted aggregate index has two disadvantages: the normal usage quantities Q_{it} must be redetermined each year, thus adding to the time and cost of data collection, and each year the index numbers for previous years must be recomputed to reflect the effect of the new quantity weights. Because of these disadvantages, the Laspeyres index is more widely used. The automotive operating expense index was computed with base-period quantities; hence, it is a Laspeyres index. Had usage figures for 1998 been used, we would have had a Paasche index. Indeed, because of more fuel efficient cars, gasoline usage has decreased and a Paasche index would differ from a Laspeyres index.

EXERCISES

Methods

1. The following table reports prices and usage quantities for two items in 1989 and 2001.

	Quantity		Unit Price ($)	
Item	1989	2001	1989	2001
A	1500	1800	7.50	7.75
B	2	1	630.00	1500.00

   a. Compute price relatives for each item in 2001 using 1989 as the base period.
   b. Compute an unweighted aggregate price index for the two items in 2001 using 1989 as the base period.
   c. Compute a weighted aggregate price index for the two items using the Laspeyres method.
   d. Compute a weighted aggregate price index for the two items using the Paasche method.

2. An item with a price relative of 132 cost $10.75 in 2001. Its base year was 1990.
   a. What was the percentage increase or decrease in cost of the item over the 11-year period?
   b. What did the item cost in 1990?

Applications

3. A large manufacturer purchases an identical component from three independent suppliers that differ in unit price and quantity supplied. The relevant data for 1999 and 2001 are given here.

		Unit Price ($)	
Supplier	Quantity (1999)	1999	2001
A	150	5.45	6.00
B	200	5.60	5.95
C	120	5.50	6.20

   a. Compute the price relatives for each of the component suppliers separately. Compare the price increases by the suppliers over the 2-year period.
   b. Compute an unweighted aggregate price index for the component part in 2001.
   c. Compute a 2001 weighted aggregate price index for the component part. What is the interpretation of this index for the manufacturing firm?

4. R&B Beverages, Inc., provides a complete line of beer, wine, and soft drink products for distribution through retail outlets in central Iowa. Unit price data for 1997 and 2001 and quantities sold in cases for 1997 follow.

	1997 Quantity	Unit Price ($)	
Item	(cases)	1997	2001
Beer	35,000	15.00	16.25
Wine	5,000	60.00	64.00
Soft drink	60,000	9.80	10.00

Compute a weighted aggregate index for the R&B Beverage sales in 2001, with 1997 as the base period.

5. Under the LIFO inventory valuation method, a price index for inventory must be established for tax purposes. The quantity weights are based on year-ending inventory levels. Use the beginning-of-the-year price per unit as the base-period price and develop a weighted aggregate index for the total inventory value at the end of the year. What type of weighted aggregate price index must be developed for the LIFO inventory valuation?

Product	Ending Inventory	Unit Price ($)	
		Beginning	**Ending**
A	500	.15	.19
B	50	1.60	1.80
C	100	4.50	4.20
D	40	12.00	13.20

17.3   COMPUTING AN AGGREGATE PRICE INDEX FROM PRICE RELATIVES

In Section 17.1 we defined the concept of a price relative and showed how a price relative can be computed with knowledge of the current-period unit price and the base-period unit price. We now want to show how aggregate price indexes like the ones developed in Section 17.2 can be computed directly from information about the price relative of each item in the group. Because of the limited use of unweighted indexes, we restrict our attention to weighted aggregate price indexes. Let us return to the automotive operating expense index of the preceding section. The necessary information for the four items is given in Table 17.5.

One must be sure prices and quantities are in the same units. For example, if prices are per case, quantity must be the number of cases and not, for instance, the number of individual units.

Let w_i be the weight applied to the price relative for item i. The general expression for a weighted average of price relatives is given by

$$I_t = \frac{\sum \frac{P_{it}}{P_{i0}}(100)w_i}{\sum w_i} \tag{17.6}$$

The proper choice of weights in (17.6) will enable us to compute a weighted aggregate price index from the price relatives. The proper choice of weights is given by multiplying the base-period price by the quantity of usage.

$$w_i = P_{i0}Q_i \tag{17.7}$$

TABLE 17.5   PRICE RELATIVES FOR AUTOMOTIVE OPERATING EXPENSE INDEX

Item	Unit Price ($)		Price Relative $(P_t/P_0)100$	Annual Usage
	1984 (P_0)	**1998** (P_t)		
Gallon of gasoline	1.21	1.06	87.6	1000
Quart of oil	1.50	2.20	146.7	15
Tires	80.00	145.00	181.3	2
Insurance policy	300.00	700.00	233.3	1

Substituting $w_i = P_{i0}Q_i$ into equation (17.6) provides the following expression for a weighted price relatives index.

$$I_t = \frac{\sum \dfrac{P_{it}}{P_{i0}}(100)(P_{i0}Q_i)}{\sum P_{i0}Q_i} \tag{17.8}$$

With the canceling of the P_{i0} terms in the numerator, an equivalent expression for the weighted price relatives index is

$$I_t = \frac{\sum P_{it}Q_i}{\sum P_{i0}Q_i}(100)$$

Thus, we see that the weighted price relatives index with $w_i = P_{i0}Q_i$ provides a price index identical to the weighted aggregate index presented in Section 17.2 by equation (17.3). Use of base-period quantities (i.e., $Q_i = Q_{i0}$) in equation (17.7) leads to a Laspeyres index. Use of current-period quantities (i.e., $Q_i = Q_{it}$) in equation (17.7) leads to a Paasche index.

Let us return to the automotive operating expense data. We can use the price relatives in Table 17.5 and equation (17.6) to compute a weighted average of price relatives. The results obtained by using the weights specified by equation (17.7) are reported in Table 17.6. The index number 123 represents a 23% increase in automotive operating expenses, which is the same as the increase identified by the weighted aggregate index computation in Section 17.2.

TABLE 17.6   AUTOMOTIVE OPERATING EXPENSE INDEX (1984–1998) BASED ON WEIGHTED PRICE RELATIVES

Item	Price Relatives $(P_{it}/P_{i0})(100)$	Base Price (\$) P_{i0}	Quantity Q_i	Weight $w_i = P_{i0}Q_i$	Weighted Price Relatives $(P_{it}/P_{i0})(100)w_i$
Gasoline	87.6	1.21	1000	1,210.0	105,996.00
Oil	146.7	1.50	15	22.5	3,300.75
Tires	181.3	80.00	2	160.0	29,008.00
Insurance	233.3	300.00	1	300.0	69,990.00
			Totals	1,692.5	208,294.75

$$I_{1998} = \frac{208,294.75}{1,692.5} = 123$$

EXERCISES

Methods

6. Price relatives for three items, along with base-period prices and usage are shown in the following table. Compute a weighted aggregate price index for the current period.

Item	Price Relative	Base Period Price	Usage
A	150	22.00	20
B	90	5.00	50
C	120	14.00	40

Applications

7. The Mitchell Chemical Company produces a special industrial chemical that is a blend of three chemical ingredients. The beginning-year cost per pound, the ending-year cost per pound, and the blend proportions follow.

| | Cost per Pound ($) | | Quantity (pounds) |
Ingredient	Beginning	Ending	per 100 Pounds of Product
A	2.50	3.95	25
B	8.75	9.90	15
C	.99	.95	60

   a. Compute the price relatives for the three ingredients.
   b. Compute a weighted average of the price relatives to develop a 1-year cost index for raw materials used in the product. What is your interpretation of this index value?

8. An investment portfolio consists of four stocks. The purchase price, current price, and number of shares are reported in the following table.

Stock	Purchase Price/Share ($)	Current Price/Share ($)	Number of Shares
Holiday Trans	15.50	17.00	500
NY Electric	18.50	20.25	200
KY Gas	26.75	26.00	500
PQ Soaps	42.25	45.50	300

   Construct a weighted average of price relatives as an index of the performance of the portfolio to date. Interpret this price index.

9. Compute the price relatives for the R&B Beverages products in Exercise 4. Use a weighted average of price relatives to show that this method provides the same index as the weighted aggregate method.

17.4   SOME IMPORTANT PRICE INDEXES

We have identified the procedures used to compute price indexes for single items or groups of items. Now let us consider some price indexes that are important measures of business and economic conditions. Specifically, we will consider the Consumer Price Index, the Producer Price Index, and the Dow Jones averages.

Consumer Price Index

The CPI includes charges for services (e.g., doctor and dentist bills) and all taxes directly associated with the purchase and use of an item.

The **Consumer Price Index (CPI)**, published monthly by the U.S. Bureau of Labor Statistics, is the primary measure of the cost of living in the United States. The group of items used to develop the index consists of a *market basket* of 400 items including food, housing, clothing, transportation, and medical items. The CPI is a weighted aggregate price index with fixed weights.* The weight applied to each item in the market basket derives from a usage survey of urban families throughout the United States.

The August 2000 CPI, computed with a 1982–1984 base index of 100, was 172.7. This figure means that the cost of purchasing the market basket of goods and services had

---

*There are actually two Consumer Price Indexes. The Bureau of Labor Statistics publishes a Consumer Price Index for all urban consumers (CPI-U) and a revised Consumer Price Index for urban wage earners and clerical workers (CPI-W). The CPI-U is the one most widely quoted, and it is published regularly in *The Wall Street Journal*.

increased 72.7% since the base period 1982–1984. The 50-year time series of the CPI from 1950 to 2000 is shown in Figure 17.1. Note how the CPI measure reflects the sharp inflationary behavior of the economy in the late 1970s and early 1980s.

Producer Price Index

The PPI is designed as a measure of price changes for domestic goods; imports are not included.

The **Producer Price Index (PPI),** also published monthly by the U.S. Bureau of Labor Statistics, measures the monthly changes in prices in primary markets in the United States. The PPI is based on prices for the first transaction of each product in nonretail markets. All commodities sold in commercial transactions in these markets are represented. The survey covers raw, manufactured, and processed goods at each level of processing and includes the output of industries classified as manufacturing, agriculture, forestry, fishing, mining, gas and electricity, and public utilities. One of the common uses of this index is as a leading indicator of the future trend of consumer prices and the cost of living. An increase in the PPI reflects producer price increases that will eventually be passed on to the consumer through higher retail prices.

Weights for the various items in the PPI are based on the value of shipments. The weighted average of price relatives is calculated by the Laspeyres method. The August 2000 PPI, computed with a 1982 base index of 100, was 138.1.

Dow Jones Averages

Charles Henry Dow published his first stock average on July 3, 1884, in the Customer's Afternoon Letter. *Eleven stocks, nine of which were railroad issues, were included in the first index. An average comparable to the DJIA was first published on October 1, 1928.*

The **Dow Jones averages** are indexes designed to show price trends and movements on the New York Stock Exchange. The best known of the Dow Jones indexes is the Dow Jones Industrial Average (DJIA), which is based on common stock prices of 30 large companies. It is the sum of these stock prices divided by a number, which is revised from time to time to adjust for stock splits and switching of companies in the index. Unlike the other price in-

FIGURE 17.1   CONSUMER PRICE INDEX, 1950–2000 WITH BASE 1982–1984 = 100

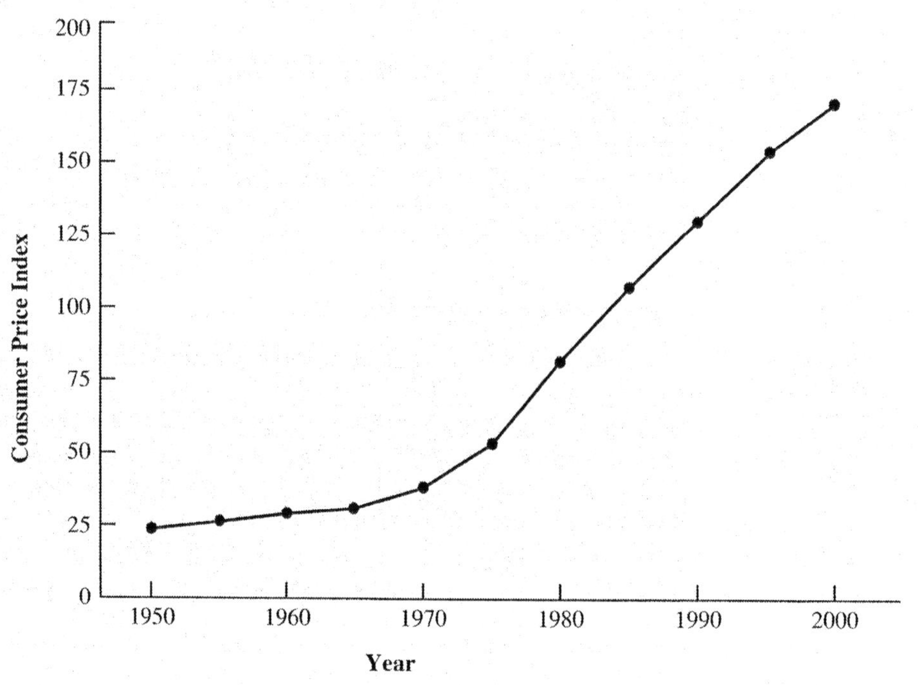

TABLE 17.7   THE 30 COMPANIES USED IN THE DOW JONES INDUSTRIAL AVERAGE (SEPTEMBER 2000)

Alcoa	Exxon Mobil	J. P. Morgan Chase
American Express	General Electric	McDonald's
AT&T	General Motors	Merck
Boeing	Hewlett-Packard	Microsoft
Caterpillar	Home Depot	Minnesota Mining
Citigroup	Honeywell Int'l	Philip Morris
Coca Cola	IBM	Procter & Gamble
Disney	Intel	SBC Communications
DuPont	International Paper	United Technologies
Eastman Kodak	Johnson & Johnson	Wal-Mart Stores

Source: Barron's, March 12, 2001.

dexes that we have studied, it is not expressed as a percentage of base-year prices. The specific firms used in September 2000 to compute the DJIA are listed in Table 17.7.

Other Dow Jones averages are computed for 20 transportation stocks and for 15 utility stocks. The Dow Jones averages are computed and published daily in *The Wall Street Journal* and other financial publications.

17.5   DEFLATING A SERIES BY PRICE INDEXES

Time series are deflated to remove the effects of inflation.

Many business and economic series reported over time, such as company sales, industry sales, and inventories, are measured in dollar amounts. These time series often show an increasing growth pattern over time, which is generally interpreted as indicating an increase in the physical volume associated with the activities. For example, a total dollar amount of inventory up by 10% might be interpreted to mean that the physical inventory is 10% larger. Such interpretations can be misleading if a time series is measured in terms of dollars, and the total dollar amount is a combination of both price and quantity changes. Hence, in periods when price changes are significant, the changes in the dollar amounts may not be indicative of quantity changes unless we are able to adjust the time series to eliminate the price change effect.

For example, from 1976 to 1980, the total amount of spending in the construction industry increased approximately 75%. That figure suggests excellent growth in construction activity. However, construction prices were increasing just as fast as—or sometimes even faster than—the 75% rate. In fact, while total construction spending was increasing, construction activity was staying relatively constant or, as in the case of new housing starts, decreasing. To interpret construction activity correctly for the 1976–1980 period, we must adjust the total spending series by a price index to remove the price increase effect. Whenever we remove the price increase effect from a time series, we say we are *deflating the series.*

In relation to personal income and wages, we often hear discussions about issues such as "real wages" or the "purchasing power" of wages. These concepts are based on the notion of deflating an hourly wage index. For example, Figure 17.2 shows the pattern of hourly wages of manufacturing workers for the period 1996–2000. We see a trend of wage increases from $12.77 per hour to $14.36 per hour. Should manufacturing workers be pleased with this growth in hourly wages? The answer depends on what has happened to the purchasing power of their wages. If we can compare the purchasing power of the $12.77 hourly wage in 1996 with the purchasing power of the $14.36 hourly wage in 2000, we will be better able to judge the relative improvement in wages.

FIGURE 17.2   ACTUAL HOURLY WAGES OF MANUFACTURING WORKERS

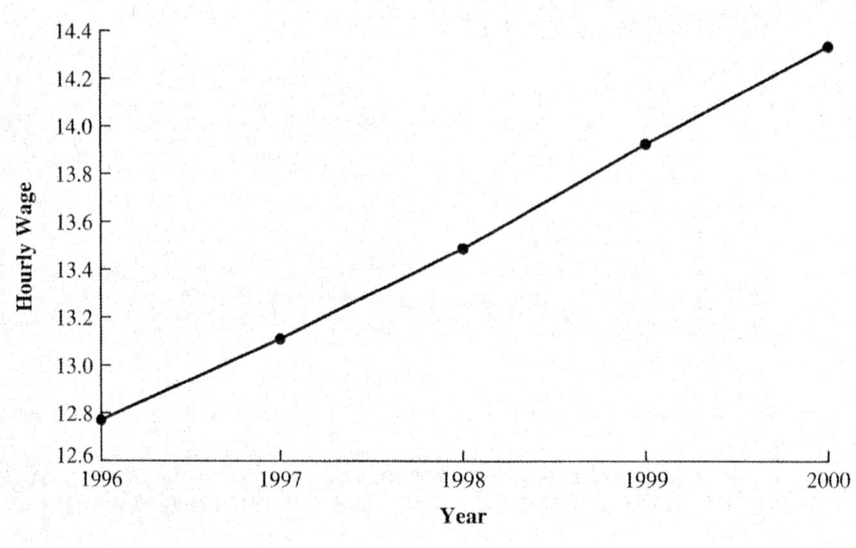

Table 17.8 reports both the hourly wage rate and the CPI for the period 1996–2000. With these data, we will show how the CPI can be used to deflate the index of hourly wages. The deflated series is found by dividing the hourly wage rate in each year by the corresponding value of the CPI and multiplying by 100. The deflated hourly wage index for manufacturing workers is given in Table 17.9; Figure 17.3 is a graph showing the deflated, or real, wages.

What does the deflated series of wages tell us about the real wages or purchasing power of workers during the 1996–2000 period? In terms of base period dollars (1982–1984 = 100), the hourly wage rate remained relatively flat over the period. After removing the inflationary effect we see that the purchasing power of the workers has changed only slightly. This effect is seen clearly in Figure 17.3. Thus, the advantage of using price indexes to deflate a series is that we have a clearer picture of the real dollar changes that are occurring.

This process of deflating a series measured over time has an important application in the computation of the Gross Domestic Product (GDP). The GDP is the total value of all goods and services produced in a given country. Obviously, over time the GDP will show gains that are in part due to price increases if the GDP is not deflated by a price index. There-

Real wages are a better measure of purchasing power than actual wages. Indeed, many union contracts call for wages to be adjusted in accordance with changes in the cost of living.

TABLE 17.8   HOURLY WAGES OF MANUFACTURING WORKERS AND CONSUMER PRICE INDEX: 1996–2000

Year	Hourly Wage ($)	CPI (1982–1984 Base)
1996	12.77	156.9
1997	13.11	160.5
1998	13.46	163.0
1999	13.93	166.6
2000	14.36	172.6

Source: Bureau of Labor Statistics

TABLE 17.9   DEFLATED SERIES OF HOURLY WAGES FOR MANUFACTURING WORKERS

Year	Deflated Hourly Wage
1996	($12.77/156.9)(100) = $8.14
1997	($13.11/160.5)(100) = $8.17
1998	($13.46/163.0)(100) = $8.26
1999	($13.93/166.6)(100) = $8.36
2000	($14.36/172.6)(100) = $8.32

FIGURE 17.3   REAL HOURLY WAGES OF MANUFACTURING WORKERS
(1982–1984 = 100)

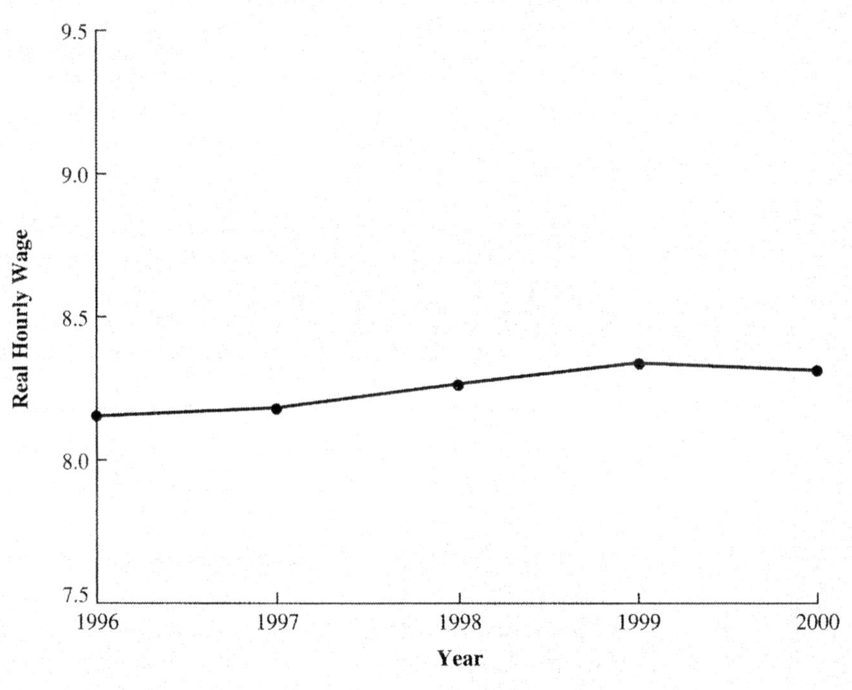

fore, to adjust the total value of goods and services to reflect actual changes in the volume of goods and services produced and sold, the GDP must be computed with a price index deflator. The process is similar to that discussed in the real wages computation.

EXERCISES

Applications

10. Average hourly wages for manufacturing workers in 1980 were $7.27; in 2000, they were $14.36. The CPI in 1980 was 82.4; in 2000 it was 172.6.
   a. Deflate the hourly wage rates in 1980 and 2000 to find the real wage rates.
   b. What is the percentage change in actual hourly wages from 1980 to 2000?
   c. What is the percentage change in real wages from 1980 to 2000?

11. Average hourly wages for workers in service industries for the 5 years from 1996 through 2000 are reported here. Use the Consumer Price Index information in Table 17.8 to deflate the wages series. What has been the percentage increase in real wages and salaries from 1998 to 2000?

Year	Total Wages & Salaries ($ billions)
1996	11.76
1997	12.23
1998	12.84
1999	13.35
2000	13.82

Source: Bureau of Labor Statistics

12. The U.S. Bureau of the Census reported the following total manufacturing shipments for the 3 years from 1997 through 1999.

Year	Manufacturing Shipments ($ billions)
1997	3929
1998	4052
1999	4260

a. The CPI for 1997–1999 is given in Table 17.8. Use this information to deflate the manufacturing shipments series and comment on the pattern of manufacturers' shipments in terms of constant dollars.

b. The following Producer Price Indexes (finished consumer goods) are for 1997 through 1999, with 1982 as the base year. Use the PPI to deflate the series.

Year	PPI (1982 = 100)
1997	131.8
1998	130.7
1999	133.0

c. Do you feel that the CPI or the PPI is more appropriate to use as a deflator for manufacturing shipments?

13. Dooley Retail Outlets' total retail sales volumes for selected years since 1982 is shown in the following table. Also shown is the CPI with the index base of 1982–1984. Deflate the sales volume figures on the basis of 1982–1984 constant dollars, and comment on the firm's sales volumes in terms of deflated dollars.

Year	Retail Sales ($)	CPI (1982–1984 base)
1982	380,000	96.5
1987	520,000	113.6
1992	700,000	140.3
1997	870,000	160.5
2000	940,000	172.6

17.6 PRICE INDEXES: OTHER CONSIDERATIONS

In the preceding sections we described several methods used to compute price indexes, discussed the use of some important indexes, and presented a procedure for using price indexes to deflate a time series. Several other issues must be considered to enhance our understanding of how price indexes are constructed and how they are used. Some are discussed in this section.

Selection of Items

The primary purpose of a price index is to measure the price change over time for a specified class of items, products, and so on. Whenever the class of items is very large, the index cannot be based on all items in the class. Rather, a sample of representative items must be used. By collecting price and quantity information for the sampled items, we hope to obtain a good idea of the price behavior of all items that the index is representing. For example, in the Consumer Price Index the total number of items that might be considered in the population of normal purchase items for a consumer could be 2000 or more. However, the index is based on the price-quantity characteristics of just 400 items. The selection of the specific items in the index is not a trivial task. Surveys of user purchase patterns as well as good judgment go into the selection process. A simple random sample is not used to select the 400 items.

After the initial selection process, the group of items in the index must be periodically reviewed and revised whenever purchase patterns change. Thus, the issue of which items to include in an index must be resolved before an index can be developed and again before it is revised.

Selection of a Base Period

Most indexes are established with a base-period value of 100 at some specific time. All future values of the index are then related to the base-period value. But what base period is appropriate for an index? It is not an easy question, and the answer must be based on the judgment of the developer of the index.

Many of the indexes established by the United States government as of 2000 have a 1982 base period. As a general guideline, the base period should not be too far from the current period. For example, a Consumer Price Index with a 1945 base period would be difficult for most individuals to understand because of unfamiliarity with conditions in 1945. The base period for most indexes therefore is adjusted periodically to a more recent period of time. The CPI base period was changed from 1967 to the 1982–1984 average in 1988. The PPI currently uses 1982 as its base period (i.e., 1982 = 100).

Quality Changes

The purpose of a price index is to measure changes in prices over time. Ideally, price data are collected for the same set of items at several times, and then the index is computed. A basic assumption is that the prices are identified for the same items each period. A problem is encountered when a product changes in quality from one period to the next. For example, a manufacturer may alter the quality of a product by using less expensive materials, fewer features, and so on, from year to year. The price may go up in following years, but the price is for a lower quality product. Consequently, the price may actually go up more than is represented by the list price for the item. It is difficult, if not impossible, to adjust an index for decreases in the quality of an item.

A substantial quality improvement also may cause an increase in the price of a product. A portion of the price related to the quality improvement should be excluded from the index computation. However, adjusting an index for a price increase that is related to higher quality of an item is extremely difficult, if not impossible.

Although common practice is to ignore minor quality changes in developing a price index, major quality changes must be addressed because they can alter the product description from period to period. If a product description is changed, the index must be modified to account for it; in some cases, the product might be deleted from the index.

In some situations, however, a substantial improvement in quality is followed by a decrease in the price. This less typical situation has been the case with personal computers during the 1990s. Designers of the CPI are now making the proper adjustments in the CPI to reflect this situation.

17.7   QUANTITY INDEXES

In addition to the price indexes described in the preceding sections, other types of indexes are useful. In particular, one other application of index numbers is to measure changes in quantity levels over time. This type of index is called a **quantity index.**

Recall that in the development of the weighted aggregate price index in Section 17.2, to compute an index number for period t we needed data on unit prices at a base period (P_0) and period t (P_t). Equation (17.3) provided the weighted aggregate price index as

$$I_t = \frac{\Sigma P_{it} Q_i}{\Sigma P_{i0} Q_i} (100)$$

The numerator, $\Sigma P_{it} Q_i$, represents the total value of fixed quantities of the index items in period t. The denominator, $\Sigma P_{i0} Q_i$, represents the total value of the same fixed quantities of the index items in year 0.

Computation of a weighted aggregate quantity index is similar to that of a weighted aggregate price index. Quantities for each item are measured in the base period and period t, with Q_{i0} and Q_{it}, respectively, representing those quantities for item i. The quantities are then weighted by a fixed price, the value added, or some other factor. The "value added" to a product is the sales value minus the cost of purchased inputs. The formula for computing a weighted aggregate quantity index for period t is

$$I_t = \frac{\Sigma Q_{it} w_i}{\Sigma Q_{i0} w_i} (100) \tag{17.9}$$

In some quantity indexes the weight for item i is taken to be the base-period price (P_{i0}), in which case the weighted aggregate quantity index is

$$I_t = \frac{\Sigma Q_{it} P_{i0}}{\Sigma Q_{i0} P_{i0}} (100) \tag{17.10}$$

Quantity indexes can also be computed on the basis of weighted quantity relatives. One formula for this version of a quantity index follows.

$$I_t = \frac{\Sigma \dfrac{Q_{it}}{Q_{i0}} (Q_{i0} P_i)}{\Sigma Q_{i0} P_i} (100) \tag{17.11}$$

This formula is the quantity version of the weighted price relatives formula developed in Section 17.3 as in equation (17.8).

The **Index of Industrial Production,** developed by the Federal Reserve Board, is probably the best-known quantity index. It is reported monthly and the base period is 1992. The index is designed to measure changes in volume of production levels for a variety of manufacturing classifications in addition to mining and utilities. In August 2000 the index was 145.7.

SELF test

EXERCISES

Methods

14. Data on quantities of three items sold in 1992 and 2000 are given here along with the sales prices of the items in 1992. Compute a weighted aggregate quantity index for 2000.

	Quantity Sold		
Item	1992	2000	Price/Unit 1992 ($)
A	350	300	18.00
B	220	400	4.90
C	730	850	15.00

Applications

SELF test

15. A trucking firm handles four commodities for a particular distributor. Total shipments for the commodities in 1991 and 2000, as well as the 1991 prices, are reported in the following table.

	Shipments		Price/Shipment
Commodity	1991	2000	1991
A	120	95	$1200
B	86	75	$1800
C	35	50	$2000
D	60	70	$1500

Develop a weighted aggregate quantity index with a 1991 base. Comment on the growth or decline in quantities over the 1991–2000 period.

16. An automobile dealer reports the 1989 and 2000 sales for three models in the following table. Compute quantity relatives and use them to develop a weighted aggregate quantity index for 2000 using the 2 years of data.

	Sales		Mean Price per Sale
Model	1989	2000	(1989)
Sedan	200	170	$15,200
Sport	100	80	$17,000
Wagon	75	60	$16,800

SUMMARY

Price and quantity indexes are important measures of changes in price and quantity levels within the business and economic environment. Price relatives are simply the ratio of the current unit price of an item to a base-period unit price multiplied by 100, with a value of 100 indicating no difference in the current- and base-period prices. Aggregate price indexes are created as a composite measure of the overall change in prices for a given group of items or products. Usually the items in an aggregate price index are weighted by their quantity of usage. A weighted aggregate price index can also be computed by weighting the price relatives by the usage quantities for the items in the index.

The Consumer Price Index and the Producer Price Index are both widely quoted indexes with 1982–1984 and 1982, respectively, as base years. The Dow Jones Industrial Average is another widely quoted price index. It is a weighted sum of the prices of 30 common stocks listed on the New York Stock Exchange. Unlike many other indexes, it is not stated as a percentage of some base-period value.

Often price indexes are used to deflate some other economic series reported over time. We saw how the CPI could be used to deflate hourly wages to obtain an index of real wages. Selection of the items to be included in the index, selection of a base period for the index, and adjustment for changes in quality are important additional considerations in the development of an index number. Quantity indexes were briefly discussed, and the Index of Industrial Production was mentioned as an important quantity index.

GLOSSARY

Price relative   A price index for a given item that is computed by dividing a current unit price by a base-period unit price and multiplying the result by 100.

Aggregate price index   A composite price index based on the prices of a group of items.

Weighted aggregate price index   A composite price index in which the prices of the items in the composite are weighted by their relative importance.

Laspeyres index   A weighted aggregate price index in which the weight for each item is its base-period quantity.

Paasche index   A weighted aggregate price index in which the weight for each item is its current-period quantity.

Consumer Price Index   A monthly price index that uses the price changes in a market basket of consumer goods and services to measure the changes in consumer prices over time.

Producer Price Index   A monthly price index designed to measure changes in prices of goods sold in primary markets (i.e., first purchase of a commodity in nonretail markets).

Dow Jones averages   Aggregate price indexes designed to show price trends and movements on the New York Stock Exchange.

Quantity index   An index that is designed to measure changes in quantities over time.

Index of Industrial Production   A quantity index that is designed to measure changes in the physical volume or production levels of industrial goods over time.

KEY FORMULAS

Price Relative in Period t

$$\frac{\text{Price in period } t}{\text{Base period price}}(100) \tag{17.1}$$

Unweighted Aggregate Price Index in Period t

$$I_t = \frac{\Sigma P_{it}}{\Sigma P_{i0}}(100) \tag{17.2}$$

Weighted Aggregate Price Index in Period t

$$I_t = \frac{\Sigma P_{it}Q_i}{\Sigma P_{i0}Q_i}(100) \tag{17.3}$$

Weighted Average of Price Relatives

$$I_t = \frac{\sum \dfrac{P_{it}}{P_{i0}}(100)w_i}{\sum w_i} \tag{17.6}$$

Weighting Factor for (17.6)

$$w_i = P_{i0}Q_i \tag{17.7}$$

Weighted Aggregate Quantity Index

$$I_t = \frac{\sum Q_{it}w_i}{\sum Q_{i0}w_i}(100) \tag{17.9}$$

SUPPLEMENTARY EXERCISES

17. The median sales prices for new single-family houses for the years 1996–1999 are as follows (*Statistical Abstract of the United States, 2000*).

Year	Price ($1000s)
1996	140.0
1997	146.0
1998	152.5
1999	160.0

   a. Use 1996 as the base year and develop a price index for new single-family homes over this 4-year period.
   b. Use 1997 as the base year and develop a price index for new single-family homes over this 4-year period.

18. Nickerson Manufacturing Company has the following data on quantities shipped and unit costs for each of its four products:

Products	Base-Period Quantities (1998)	Mean Shipping Cost per Unit ($) 1998	Mean Shipping Cost per Unit ($) 2001
A	2000	10.50	15.90
B	5000	16.25	32.00
C	6500	12.20	17.40
D	2500	20.00	35.50

   a. Compute the price relative for each product.
   b. Compute a weighted aggregate price index that reflects the shipping cost change over the four-year period.

19. Use the price data in Exercise 18 to compute a Paasche index for the shipping cost if 2001 quantities are 4000, 3000, 7500, and 3000 for each of the four products.

20. Boran Stockbrokers, Inc., selects four stocks for the purpose of developing its own index of stock market behavior. Costs per share for a 1999 base period, January 2001, and March 2001 follow. Base-year quantities have been set on the basis of historical volumes for the four stocks.

| | | | | Cost per Share ($) | |
Stock	Industry	1999 Quantity	1999 Base	January 2001	March 2001
A	Oil	100	31.50	32.75	32.50
B	Computer	150	65.00	59.00	57.50
C	Steel	75	40.00	42.00	39.50
D	Real Estate	50	18.00	16.50	13.75

Use the 1999 base period to compute the Boran index for January 2001 and March 2001. Comment on what the index tells you about what is happening in the stock market.

21. Compute the price relatives for the four stocks making up the Boran index in Exercise 20. Use the weighted aggregates of price relatives to compute the January 2001 and March 2001 Boran indexes.

22. Consider the following price relatives and quantity information for grain production in Iowa (*Statistical Abstract of the United States,* 1997).

Product	1991 Quantities (millions of bushels)	Base Price per Bushel ($)	1991–1996 Price Relatives
Corn	1427	2.30	113
Soybeans	350	5.51	123

What is the 1996 weighted aggregate price index for the Iowa grains?

23. Fresh fruit price and quantity data for the years 1988 and 1998 follow (*Statistical Abstract of the United States,* 1999). Quantity data reflect per capita consumption in pounds and prices are per pound.

Fruit	1988 per Capita Consumption (pounds)	1988 Price ($/pound)	1998 Price ($/pound)
Bananas	24.3	.41	.51
Apples	19.9	.71	.85
Oranges	13.9	.56	.61
Pears	3.2	.64	.98

a. Compute a price relative for each product.
b. Compute a weighted aggregate price index for fruit products. Comment on the change in fruit prices over the 10-year period.

24. Starting faculty salaries (9-month basis) for assistant professors of business administration at a major Midwestern university follow. Use the CPI to deflate the salary data to constant dollars. Comment on the trend in salaries in higher education as indicated by these data.

Year	Starting Salary ($)	CPI (1982–1984 Base)
1970	14,000	38.8
1975	17,500	53.8
1980	23,000	82.4
1985	37,000	107.6
1990	53,000	130.7
1995	65,000	152.4
2000	80,000	172.6

25. The 5-year historical prices per share for a particular stock and the Consumer Price Index with a 1982–1984 base period follow.

Year	Price per Share ($)	CPI (1982–1984 Base)
1996	51.00	156.9
1997	54.00	160.5
1998	58.00	163.0
1999	59.50	166.6
2000	59.00	172.6

Deflate the stock price series and comment on the investment aspects of this stock.

26. A major manufacturing company has reported the quantity and product value information for 1997 and 2001 in the table that follows. Compute a weighted aggregate quantity index for the data. Comment on what this quantity index means.

Product	Quantities		Values ($)
	1997	**2001**	
A	800	1200	30.00
B	600	500	20.00
C	200	500	25.00

FORECASTING

CONTENTS

STATISTICS IN PRACTICE

NEVADA OCCUPATIONAL HEALTH CLINIC*
Sparks, Nevada

Nevada Occupational Health Clinic is a privately owned medical clinic in Sparks, Nevada. The clinic specializes in industrial medicine and has been in operation at the same site for more than 20 years. In the beginning of 1991, the clinic entered a rapid growth phase in which monthly billings increased from $57,000 to more than $300,000 in 26 months. The clinic was still undergoing dramatic growth when the main clinic building burned to the ground on April 6, 1993.

The clinic's insurance policy covered physical property and equipment as well as loss of income due to the interruption of regular business operations. Settling the property insurance claim was a relatively straightforward matter of determining the value of the physical property and equipment lost during the fire. However, determining the value of the income lost during the seven months that it took to rebuild the clinic was a complicated matter involving negotiations between the business owners and the insurance company. No preestablished rules could help calculate "what would have happened" to the clinic's

A 1993 fire closed the Nevada Occupational Health Clinic for seven months. © PhotoDisc, Inc.

billings if the fire had not occurred. To estimate the lost income, the clinic used a forecasting method to project the growth in business that would have been realized during the seven-month lost-business period. The actual history of billings prior to the fire provided the basis for a forecasting model with linear trend and seasonal components as discussed in this chapter. This forecasting model enabled the clinic to establish an accurate estimate of the loss, which eventually was accepted by the insurance company.

*The authors are indebted to Bard Betz, Director of Operations, and Curtis Brauer, Executive Administrative Assistant, Nevada Occupational Health Clinic, for providing this Statistics in Practice.

An essential aspect of managing any organization is planning for the future. Indeed, the long-run success of an organization is closely related to how well management is able to anticipate the future and develop appropriate strategies. Good judgment, intuition, and an awareness of the state of the economy may give a manager a rough idea or "feeling" of what is likely to happen in the future. However, converting that feeling into a number that can be used as next quarter's sales volume or next year's raw material cost is difficult. The purpose of this chapter is to introduce several forecasting methods.

Most companies can forecast total demand for all products with errors of less than 5%. However, forecasting demand for individual products can result in significantly higher errors.

Suppose we have been asked to provide quarterly forecasts of the sales volume for a particular product during the coming one-year period. Production schedules, raw material purchasing, inventory policies, and sales quotas will all be affected by the quarterly forecasts we provide. Consequently, poor forecasts may result in poor planning and hence increased costs for the firm. How should we go about providing the quarterly sales volume forecasts?

We will certainly want to review the actual sales data for the product in past periods. Using these historical data, we can identify the general level of sales and any trend such as an increase or decrease in sales volume over time. A further review of the data might reveal a seasonal pattern such as peak sales occurring in the third quarter of each year and sales volume bottoming out during the first quarter. By reviewing historical data, we can often develop a better understanding of the pattern of past sales, leading to better predictions of future sales for the product.

A forecast is simply a prediction of what will happen in the future. Managers must learn to accept the fact that regardless of the technique used, they will not be able to develop perfect forecasts.

Historical sales form a time series. A **time series** is a set of observations on a variable measured at successive points in time or over successive periods of time. In this chapter, we will introduce several procedures for analyzing time series. The objective of such analyses is to provide good **forecasts** or predictions of future values of the time series.

Forecasting methods can be classified as quantitative or qualitative. Quantitative forecasting methods can be used when (1) past information about the variable being forecast is available, (2) the information can be quantified, and (3) a reasonable assumption is that the pattern of the past will continue into the future. In such cases, a forecast can be developed using a time series method or a causal method.

If the historical data are restricted to past values of the variable, the forecasting procedure is called a *time series method*. The objective of time series methods is to discover a pattern in the historical data and then extrapolate the pattern into the future; the forecast is based solely on past values of the variable and/or on past forecast errors. In this chapter we discuss three time series methods: smoothing (moving averages, weighted moving averages, and exponential smoothing), trend projection, and trend projection adjusted for seasonal influence.

Causal forecasting methods are based on the assumption that the variable we are forecasting has a cause-effect relationship with one or more other variables. In this chapter we discuss the use of regression analysis as a causal forecasting method. For instance, the sales volume for many products is influenced by advertising expenditures, so regression analysis may be used to develop an equation showing how these two variables are related. Then, once the advertising budget has been set for the next period, we could substitute this value into the equation to develop a prediction or forecast of the sales volume for that period. Note that if a time series method had been used to develop the forecast, advertising expenditures would not be considered; that is, a time series method would have based the forecast solely on past sales.

Qualitative methods generally involve the use of expert judgment to develop forecasts. For instance, a panel of experts might develop a consensus forecast of the prime rate for a year from now. An advantage of qualitative procedures is that they can be applied when the information on the variable being forecast cannot be quantified and when historical data are either not applicable or unavailable. Figure 18.1 provides an overview of the types of forecasting methods.

18.1   COMPONENTS OF A TIME SERIES

The pattern or behavior of the data in a time series has several components. The usual assumption is that four separate components—trend, cyclical, seasonal, and irregular—combine to provide specific values for the time series. Let us look more closely at each of these components.

Trend Component

In time series analysis, the measurements may be taken every hour, day, week, month, or year, or at any other regular interval.* Although time series data generally exhibit random fluctuations, the time series may still show gradual shifts or movements to relatively higher or lower values over a longer period of time. The gradual shifting of the time series is re-

---

*We limit our discussion to time series in which the values of the series are recorded at equal intervals. Cases in which the observations are not made at equal intervals are beyond the scope of this text.

FIGURE 18.1   OVERVIEW OF FORECASTING METHODS

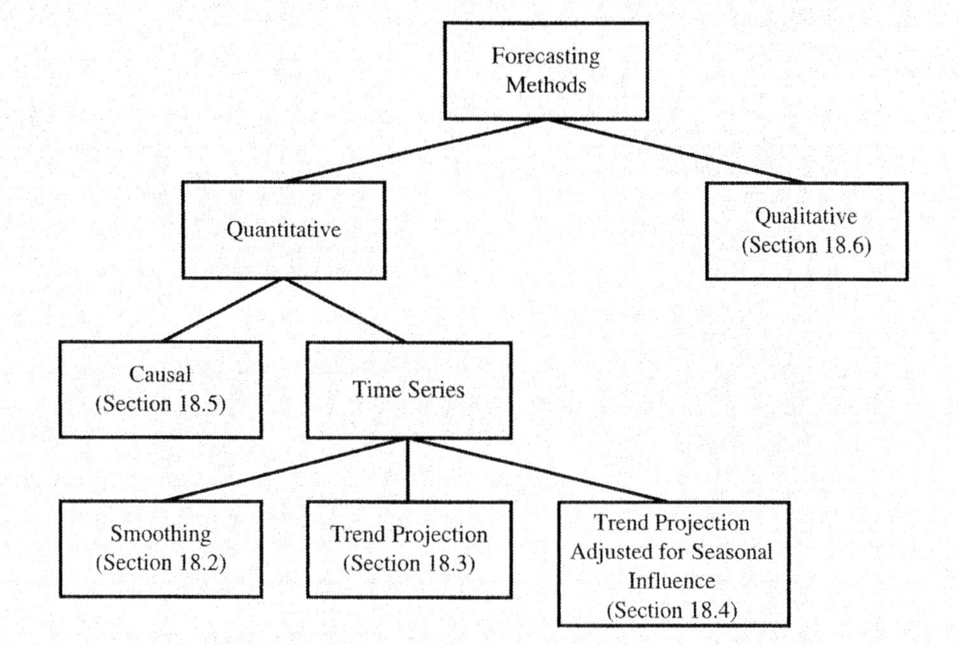

ferred to as the **trend** in the time series; this shifting or trend is usually the result of long-term factors such as changes in the population, demographic characteristics of the population, technology, and/or consumer preferences.

For example, a manufacturer of photographic equipment may see substantial month-to-month variability in the number of cameras sold. However, in reviewing the sales over the past 10 to 15 years, the manufacturer may find a gradual increase in the annual sales volume. Suppose the sales volume was approximately 17,000 cameras in 1991, 23,000 cameras in 1996, and 25,000 cameras in 2001. This gradual growth in sales over time shows an upward trend for the time series. Figure 18.2 shows a straight line that may be a good approximation of the trend in camera sales. Although the trend for camera sales appears to be linear and increasing over time, sometimes the trend in a time series can be described better by some other patterns.

Figure 18.3 shows some other possible time series trend patterns. Panel (A) shows a nonlinear trend; in this case, the time series indicates little growth initially, then a period of rapid growth, and finally a leveling off. This trend might be a good approximation of sales for a product from introduction through a growth period and into a period of market saturation. The linear decreasing trend in panel (B) is useful for a time series displaying a steady decrease over time. The horizontal line in panel (C) represents a time series that has no consistent increase or decrease over time and thus no trend.

Cyclical Component

Although a time series may exhibit a trend over long periods of time all future values of the time series will not fall exactly on the trend line. In fact, time series often show alternating sequences of points below and above the trend line. Any recurring sequence of points above

FIGURE 18.2   LINEAR TREND OF CAMERA SALES

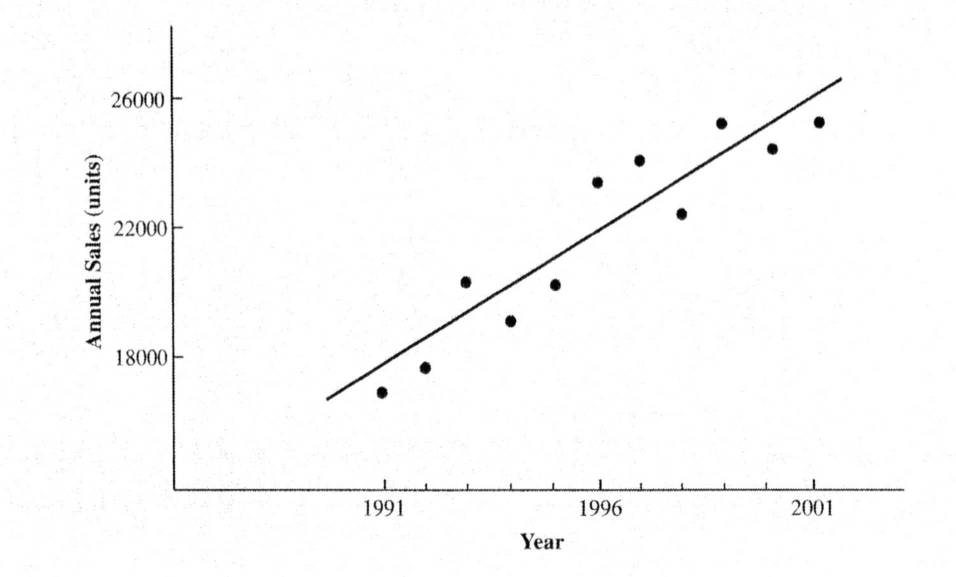

FIGURE 18.3   EXAMPLES OF SOME POSSIBLE TIME SERIES TREND PATTERNS

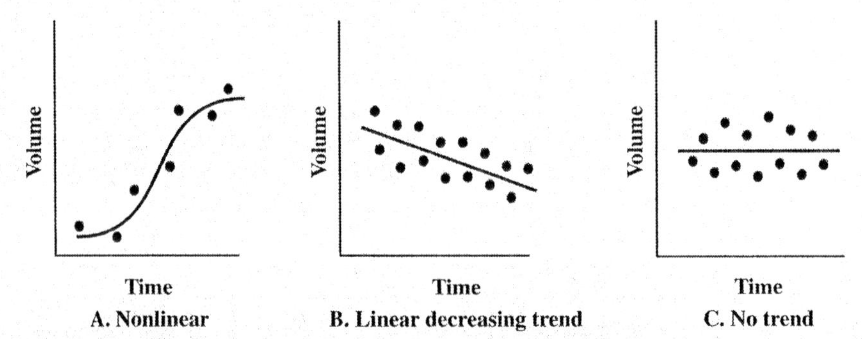

and below the trend line lasting more than one year can be attributed to the **cyclical component** of the time series. Figure 18.4 shows the graph of a time series with an obvious cyclical component. The observations are taken at intervals one year apart.

Many time series exhibit cyclical behavior with regular runs of observations below and above the trend line. Generally, this component of the time series is due to multiyear cyclical movements in the economy. For example, periods of moderate inflation followed by periods of rapid inflation can lead to time series that alternate below and above a generally increasing trend line (e.g., a time series for housing costs). Many time series in the early 1980s displayed this type of behavior.

Seasonal Component

Whereas the trend and cyclical components of a time series are identified by analyzing multiyear movements in historical data, many time series show a regular pattern over one-year periods. For example, a manufacturer of swimming pools expects low sales activity in the fall and winter months, with peak sales in the spring and summer months. Manufactur-

FIGURE 18.4   TREND AND CYCLICAL COMPONENTS OF A TIME SERIES WITH DATA
POINTS ONE YEAR APART

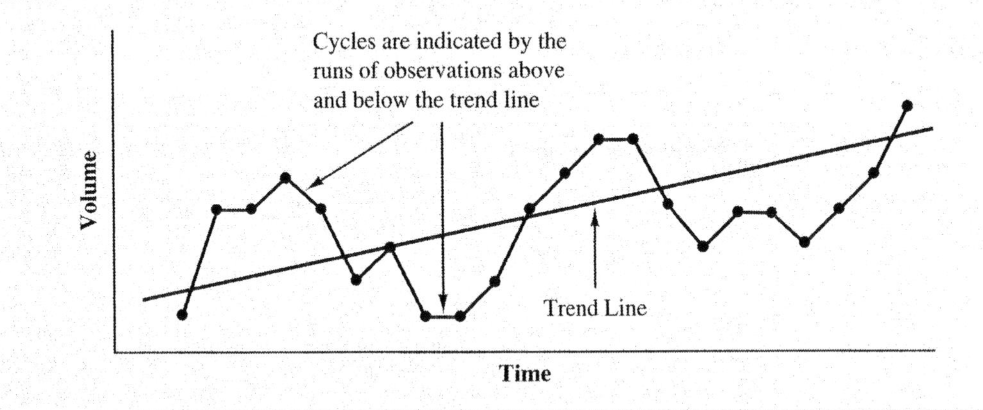

ers of snow removal equipment and heavy clothing, however, expect just the opposite
yearly pattern. Not surprisingly, the component of the time series that represents the variability in the data due to seasonal influences is called the **seasonal component.** Although
we generally think of seasonal movement in a time series as occurring within one year, the
seasonal component can also be used to represent any regularly repeating pattern that is less
than one year in duration. For example, daily traffic volume data show within-the-day "seasonal" behavior, with peak levels occurring during rush hours, moderate flow during the
rest of the day and early evening, and light flow from midnight to early morning.

Irregular Component

The **irregular component** of the time series is the residual, or "catch-all," factor that accounts for the deviations of the actual time series values from those expected given the effects of the trend, cyclical, and seasonal components. The irregular component is caused by
the short-term, unanticipated, and nonrecurring factors that affect the time series. Because
this component accounts for the random variability in the time series, it is unpredictable.
We cannot attempt to predict its impact on the time series.

18.2   SMOOTHING METHODS

*Many manufacturing
environments require
forecasts for thousands of
items weekly or monthly.
Thus, in choosing a
forecasting technique,
simplicity and ease of use
are important criteria. The
data requirements for the
techniques presented in this
section are minimal, and
the techniques are easy to
use and understand.*

In this section we discuss three forecasting methods: moving averages, weighted moving
averages, and exponential smoothing. The objective of each of these methods is to "smooth
out" the random fluctuations caused by the irregular component of the time series, therefore they are referred to as smoothing methods. Smoothing methods are appropriate for a
stable time series—that is, one that exhibits no significant trend, cyclical, or seasonal effects—because they adapt well to changes in the level of the time series. However, without
modification, they do not work as well when a significant trend, cyclical, and/or seasonal
variation is present.

Smoothing methods are easy to use and generally provide a high level of accuracy for
short-range forecasts, such as a forecast for the next time period. One of the methods, exponential smoothing, has minimal data requirements and thus is a good method to use when
forecasts are required for large numbers of items.

Moving Averages

The **moving averages** method uses the average of the most recent n data values in the time series as the forecast for the next period. Mathematically, the moving average calculation is made as follows.

Moving Average

$$\text{Moving Average} = \frac{\Sigma(\text{most recent } n \text{ data values})}{n} \qquad \textbf{(18.1)}$$

TABLE 18.1

GASOLINE SALES TIME SERIES

Week	Sales (1000s of gallons)
1	17
2	21
3	19
4	23
5	18
6	16
7	20
8	18
9	22
10	20
11	15
12	22

The term "moving" is used because every time a new observation becomes available for the time series, it replaces the oldest observation in (18.1) and a new average is computed. As a result, the average will change, or move, as new observations become available.

To illustrate the moving averages method, consider the 12 weeks of data in Table 18.1 and Figure 18.5. These data show the number of gallons of gasoline sold by a gasoline distributor in Bennington, Vermont, over the past 12 weeks. Figure 18.5 indicates that, although random variability is present, the time series appears to be stable over time. Hence, the smoothing methods of this section are applicable.

To use moving averages to forecast gasoline sales, we must first select the number of data values to be included in the moving average. As an example, let us compute forecasts using a 3-week moving average. The moving average calculation for the first 3 weeks of the gasoline sales time series is

$$\text{Moving Average (Weeks 1–3)} = \frac{17 + 21 + 19}{3} = 19$$

We then use this moving average as the forecast for week 4. Because the actual value observed in week 4 is 23, the forecast error in week 4 is $23 - 19 = 4$. In general, the error associated with any forecast is the difference between the observed value of the time series and the forecast.

The calculation for the second 3-week moving average is

$$\text{Moving Average (Weeks 2–4)} = \frac{21 + 19 + 23}{3} = 21$$

Hence, the forecast for week 5 is 21. The error associated with this forecast is $18 - 21 = -3$. Thus, the forecast error may be positive or negative depending on whether the forecast is too low or too high. A complete summary of the 3-week moving average calculations for the gasoline sales time series is provided in Table 18.2 and Figure 18.6.

Forecast Accuracy.   An important consideration in selecting a forecasting method is the accuracy of the forecast. Clearly, we want forecast errors to be small. The last two columns of Table 18.2, which contain the forecast errors and the squared forecast errors, can be used to develop a measure of accuracy.

Forecast accuracy is not the only consideration. Sometimes the most accurate method requires data on related time series that are difficult or costly to obtain. Trade-offs are often made between cost and forecast accuracy.

For the gasoline sales time series, we can use the last column of Table 18.2 to compute the average of the sum of the squared errors. Doing so we obtain

$$\text{Average of the Sum of Squared Errors} = \frac{92}{9} = 10.22$$

FIGURE 18.5   GASOLINE SALES TIME SERIES

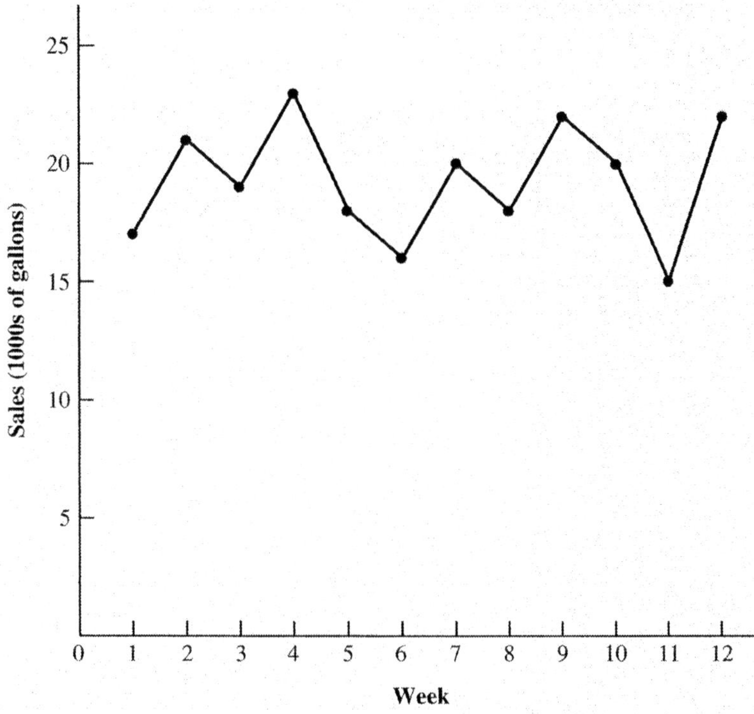

This average of the sum of squared errors is commonly referred to as the **mean squared error** (MSE). The MSE is an often-used measure of the accuracy of a forecasting method and is the one we use in this chapter.

As we indicated previously, to use the moving averages method, we must first select the number of data values to be included in the moving average. Not surprisingly, for a particular time series, moving averages of different lengths will differ in their ability to

TABLE 18.2   SUMMARY OF THREE-WEEK MOVING AVERAGE CALCULATIONS

Week	Time Series Value	Moving Average Forecast	Forecast Error	Squared Forecast Error
1	17			
2	21			
3	19			
4	23	19	4	16
5	18	21	−3	9
6	16	20	−4	16
7	20	19	1	1
8	18	18	0	0
9	22	18	4	16
10	20	20	0	0
11	15	20	−5	25
12	22	19	3	9
			Totals 0	92

FIGURE 18.6   GASOLINE SALES TIME SERIES AND THREE-WEEK MOVING AVERAGE FORECASTS

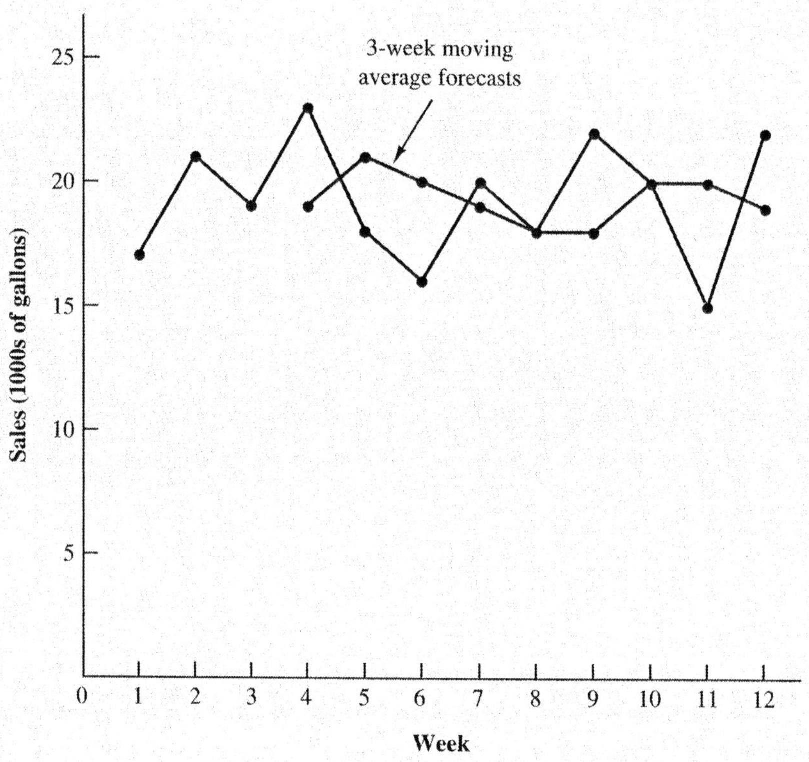

forecast the time series accurately. One possible approach to choosing the number of values to be included in the moving average is to use trial and error to identify the length that minimizes the MSE. Then, if we are willing to assume that the length that is best for the past will also be best for the future, we would forecast the next value in the time series by using the number of data values that minimized the MSE for the historical time series. Exercise 2 at the end of the section will ask you to consider 4-week and 5-week moving averages for the gasoline sales data. A comparison of the MSEs will indicate the number of weeks of data you may want to include in the moving average calculation.

Weighted Moving Averages

In the moving averages method, each observation in the moving average calculation receives the same weight. One variation, known as **weighted moving averages,** involves selecting a different weight for each data value and then computing a weighted average of the most recent n values as the forecast. In most cases, the most recent observation receives the most weight, and the weight decreases for older data values. For example, we can use the gasoline sales time series to illustrate the computation of a weighted 3-week moving average, with the most recent observation receiving a weight three times as great as that given the oldest observation, and the next oldest observation receiving a weight twice as great as the oldest. For week 4 the computation is:

$$\text{Forecast for Week 4} = \tfrac{1}{6}(17) + \tfrac{2}{6}(21) + \tfrac{3}{6}(19) = 19.33$$

Note that for the weighted moving average the sum of the weights is equal to 1. Actually the sum of the weights for the simple moving average also equalled 1: Each weight was 1/3. However, recall that the simple or unweighted moving average provided a forecast of 19.

Forecast Accuracy.   To use the weighted moving averages method we must first select the number of data values to be included in the weighted moving average and then choose weights for each of the data values. In general, if we believe that the recent past is a better predictor of the future than the distant past, larger weights should be given to the more recent observations. However, when the time series is highly variable, selecting approximately equal weights for the data values may be best. Note that the only requirement in selecting the weights is that their sum must equal 1. To determine whether one particular combination of number of data values and weights provides a more accurate forecast than another combination, we will continue to use the MSE criterion as the measure of forecast accuracy. That is, if we assume that the combination that is best for the past will also be best for the future, we would use the combination of number of data values and weights that minimized MSE for the historical time series to forecast the next value in the time series.

Exponential Smoothing

Exponential smoothing uses a weighted average of past time series values as the forecast; it is a special case of the weighted moving averages method in which we select only one weight—the weight for the most recent observation. The weights for the other data values are computed automatically and become smaller as the observations move farther into the past. The basic exponential smoothing model follows.

Exponential Smoothing Model

$$F_{t+1} = \alpha Y_t + (1 - \alpha)F_t \qquad (18.2)$$

where

$$F_{t+1} = \text{forecast of the time series for period } t + 1$$
$$Y_t = \text{actual value of the time series in period } t$$
$$F_t = \text{forecast of the time series for period } t$$
$$\alpha = \text{smoothing constant } (0 \leq \alpha \leq 1)$$

Equation (18.2) shows that the forecast for period $t + 1$ is a weighted average of the actual value in period t and the forecast for period t; note in particular that the weight given to the actual value in period t is α and that the weight given to the forecast in period t is $1 - \alpha$. We can demonstrate that the exponential smoothing forecast for any period is also a weighted average of *all the previous actual values* for the time series with a time series consisting of three periods of data: Y_1, Y_2, and Y_3. To start the calculations, we let F_1 equal the actual value of the time series in period 1; that is, $F_1 = Y_1$. Hence, the forecast for period 2 is

$$\begin{aligned} F_2 &= \alpha Y_1 + (1 - \alpha)F_1 \\ &= \alpha Y_1 + (1 - \alpha)Y_1 \\ &= Y_1 \end{aligned}$$

Thus, the exponential smoothing forecast for period 2 is equal to the actual value of the time series in period 1.

The forecast for period 3 is

$$F_3 = \alpha Y_2 + (1 - \alpha)F_2 = \alpha Y_2 + (1 - \alpha)Y_1$$

Finally, substituting this expression for F_3 in the expression for F_4, we obtain

$$\begin{aligned} F_4 &= \alpha Y_3 + (1 - \alpha)F_3 \\ &= \alpha Y_3 + (1 - \alpha)[\alpha Y_2 + (1 - \alpha)Y_1] \\ &= \alpha Y_3 + \alpha(1 - \alpha)Y_2 + (1 - \alpha)^2 Y_1 \end{aligned}$$

The term exponential smoothing *comes from the exponential nature of the weighting scheme for the historical values.*

Hence, F_4 is a weighted average of the first three time series values. The sum of the coefficients, or weights, for Y_1, Y_2, and Y_3 equals one. A similar argument can be made to show that, in general, any forecast F_{t+1} is a weighted average of all the previous time series values.

Despite the fact that exponential smoothing provides a forecast that is a weighted average of all past observations, all past data do not need to be saved to compute the forecast for the next period. In fact, once the **smoothing constant** α has been selected, only two pieces of information are needed to compute the forecast. Equation (18.2) shows that with a given α we can compute the forecast for period $t + 1$ simply by knowing the actual and forecast time series values for period t—that is, Y_t and F_t.

To illustrate the exponential smoothing approach to forecasting, consider the gasoline sales time series in Table 18.1 and Figure 18.5. As indicated, the exponential smoothing forecast for period 2 is equal to the actual value of the time series in period 1. Thus, with $Y_1 = 17$, we will set $F_2 = 17$ to start the exponential smoothing computations. Referring to the time series data in Table 18.1, we find an actual time series value in period 2 of $Y_2 = 21$. Thus, period 2 has a forecast error of $21 - 17 = 4$.

Continuing with the exponential smoothing computations using a smoothing constant of $\alpha = .2$, we obtain the following forecast for period 3.

$$F_3 = .2Y_2 + .8F_2 = .2(21) + .8(17) = 17.8$$

Once the actual time series value in period 3, $Y_3 = 19$, is known, we can generate a forecast for period 4 as follows.

$$F_4 = .2Y_3 + .8F_3 = .2(19) + .8(17.8) = 18.04$$

By continuing the exponential smoothing calculations, we can determine the weekly forecast values and the corresponding weekly forecast errors, as shown in Table 18.3. Note that we have not shown an exponential smoothing forecast or the forecast error for period 1 because no forecast was made. For week 12, we have $Y_{12} = 22$ and $F_{12} = 18.48$. Can we use this information to generate a forecast for week 13 before the actual value of week 13 becomes known? Using the exponential smoothing model, we have

$$F_{13} = .2Y_{12} + .8F_{12} = .2(22) + .8(18.48) = 19.18$$

Thus, the exponential smoothing forecast of the amount sold in week 13 is 19.18, or 19,180 gallons of gasoline. With this forecast, the firm can make plans and decisions accordingly. The accuracy of the forecast will not be known until the end of week 13.

Figure 18.7 is the plot of the actual and forecast time series values. Note in particular how the forecasts "smooth out" the irregular fluctuations in the time series.

Forecast Accuracy. In the preceding exponential smoothing calculations, we used a smoothing constant of $\alpha = .2$. Although any value of α between 0 and 1 is acceptable, some

TABLE 18.3 SUMMARY OF THE EXPONENTIAL SMOOTHING FORECASTS AND FORECAST ERRORS FOR GASOLINE SALES WITH SMOOTHING CONSTANT $\alpha = .2$

Week (t)	Time Series Value (Y_t)	Exponential Smoothing Forecast (F_t)	Forecast Error ($Y_t - F_t$)
1	17		
2	21	17.00	4.00
3	19	17.80	1.20
4	23	18.04	4.96
5	18	19.03	−1.03
6	16	18.83	−2.83
7	20	18.26	1.74
8	18	18.61	−.61
9	22	18.49	3.51
10	20	19.19	.81
11	15	19.35	−4.35
12	22	18.48	3.52

FIGURE 18.7 ACTUAL AND FORECAST GASOLINE SALES TIME SERIES WITH SMOOTHING CONSTANT $\alpha = .2$

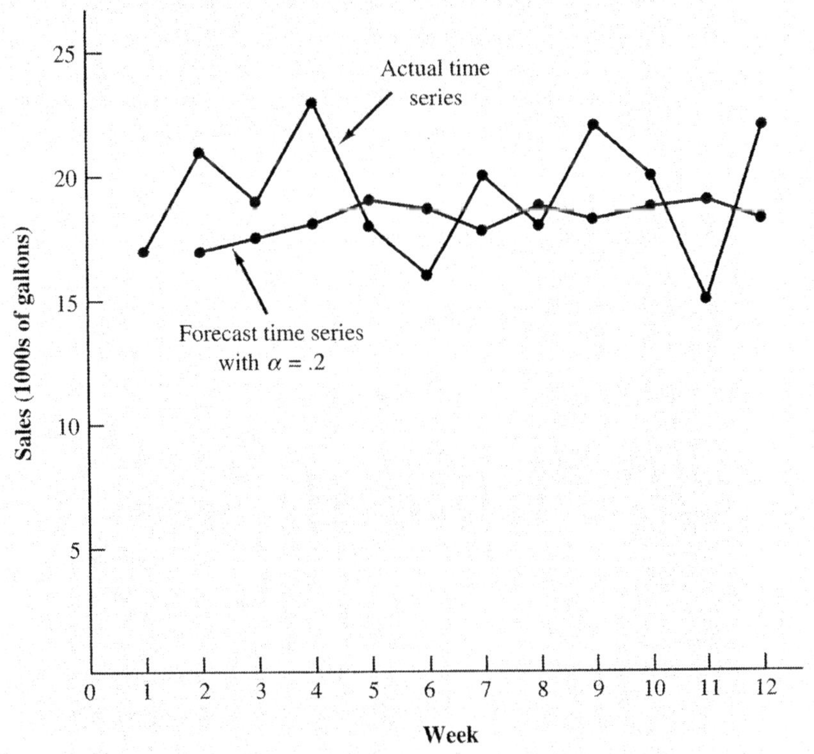

values will yield better forecasts than others. Insight into choosing a good value for α can be obtained by rewriting the basic exponential smoothing model as follows.

$$F_{t+1} = \alpha Y_t + (1 - \alpha)F_t$$
$$F_{t+1} = \alpha Y_t + F_t - \alpha F_t$$
$$F_{t+1} = \underset{\substack{\text{Forecast} \\ \text{in period } t}}{F_t} + \alpha \underset{\substack{\text{Forecast error} \\ \text{in period } t}}{(Y_t - F_t)} \tag{18.3}$$

Thus, the new forecast F_{t+1} is equal to the previous forecast F_t plus an adjustment, which is α times the most recent forecast error, $Y_t - F_t$. That is, the forecast in period $t + 1$ is obtained by adjusting the forecast in period t by a fraction of the forecast error. If the time series contains substantial random variability, a small value of the smoothing constant is preferred. The reason for this choice is that, because much of the forecast error is due to random variability, we do not want to overreact and adjust the forecasts too quickly. For a time series with relatively little random variability, larger values of the smoothing constant have the advantage of quickly adjusting the forecasts when forecasting errors occur and thus allowing the forecasts to react faster to changing conditions.

The criterion we will use to determine a desirable value for the smoothing constant α is the same as the criterion we proposed for determining the number of periods of data to include in the moving averages calculation. That is, we choose the value of α that minimizes the mean squared error (MSE). A summary of the MSE calculations for the exponential smoothing forecast of gasoline sales with $\alpha = .2$ is shown in Table 18.4. Note that there is one less squared error term than the number of time periods, because we had no past values with which to make a forecast for period 1. Would a different value of α have provided better results in terms of a lower MSE value? Perhaps the most straightforward way to answer this question is simply to try another value for α. We will then

TABLE 18.4   MSE COMPUTATIONS FOR FORECASTING GASOLINE SALES WITH $\alpha = .2$

Week (t)	Time Series Value (Y_t)	Forecast (F_t)	Forecast Error ($Y_t - F_t$)	Squared Forecast Error ($Y_t - F_t)^2$
1	17			
2	21	17.00	4.00	16.00
3	19	17.80	1.20	1.44
4	23	18.04	4.96	24.60
5	18	19.03	−1.03	1.06
6	16	18.83	−2.83	8.01
7	20	18.26	1.74	3.03
8	18	18.61	−.61	.37
9	22	18.49	3.51	12.32
10	20	19.19	.81	.66
11	15	19.35	−4.35	18.92
12	22	18.48	3.52	12.39
			Total	98.80

$$\text{MSE} = \frac{98.80}{11} = 8.98$$

TABLE 18.5   MSE COMPUTATIONS FOR FORECASTING GASOLINE SALES WITH $\alpha = .3$

Week (t)	Time Series Value (Y_t)	Forecast (F_t)	Forecast Error $(Y_t - F_t)$	Squared Forecast Error $(Y_t - F_t)^2$
1	17			
2	21	17.00	4.00	16.00
3	19	18.20	.80	.64
4	23	18.44	4.56	20.79
5	18	19.81	−1.81	3.28
6	16	19.27	−3.27	10.69
7	20	18.29	1.71	2.92
8	18	18.80	−.80	.64
9	22	18.56	3.44	11.83
10	20	19.59	.41	.17
11	15	19.71	−4.71	22.18
12	22	18.30	3.70	13.69
			Total	102.83

$$\text{MSE} = \frac{102.83}{11} = 9.35$$

compare its mean squared error with the MSE value of 8.98 obtained by using a smoothing constant of $\alpha = .2$.

The exponential smoothing results with $\alpha = .3$ are shown in Table 18.5. With MSE $= 9.35$, we see that for the current data set, a smoothing constant of $\alpha = .3$ results in less forecast accuracy than a smoothing constant of $\alpha = .2$. Thus, we would be inclined to prefer the original smoothing constant of $\alpha = .2$. Using a trial-and-error calculation with other values of α, we can find a "good" value for the smoothing constant. This value can be used in the exponential smoothing model to provide forecasts for the future. At a later date, after new time series observations have been obtained, we analyze the newly collected time series data to determine whether the smoothing constant should be revised to provide better forecasting results.

NOTES AND COMMENTS

1. Another measure of forecast accuracy is the *mean absolute deviation* (MAD). This measure is simply the average of the absolute values of all the forecast errors. Using the errors given in Table 18.2, we obtain

$$\text{MAD} = \frac{4 + 3 + 4 + 1 + 0 + 4 + 0 + 5 + 3}{9} = 2.67$$

One major difference between MSE and MAD is that the MSE measure is influenced much more by large forecast errors than by small errors (because for the MSE measure the errors are squared). The selection of the best measure of forecasting accuracy is not a simple matter. Indeed, forecasting experts often disagree as to which measure should be used. We use the MSE measure in this chapter.

2. Spreadsheet packages are an effective aid in choosing a good value of α for exponential smoothing and selecting weights for the weighted moving averages method. With the time series data and the forecasting formulas in the spreadsheets, you can experiment with different values of α (or moving average weights) and choose the value(s) of α providing the smallest MSE or MAD.

EXERCISES

Methods

1. Consider the following time series data.

Week	1	2	3	4	5	6
Value	8	13	15	17	16	9

   a. Develop a 3-week moving average for this time series. What is the forecast for week 7?
   b. Compute the MSE for the 3-week moving average.
   c. Use $\alpha = .2$ to compute the exponential smoothing values for the time series. What is the forecast for week 7?
   d. Compare the 3-week moving average forecast with the exponential smoothing forecast using $\alpha = .2$. Which appears to provide the better forecast?
   e. Use a smoothing constant of .4 to compute the exponential smoothing values. Does a smoothing constant of .2 or .4 appear to provide the better forecast? Explain.

2. Refer to the gasoline sales time series data in Table 18.1.
   a. Compute 4-week and 5-week moving averages for the time series.
   b. Compute the MSE for the 4-week and 5-week moving average forecasts.
   c. What appears to be the best number of weeks of past data to use in the moving average computation? Remember that the MSE for the 3-week moving average is 10.22.

3. Refer again to the gasoline sales time series data in Table 18.1.
   a. Using a weight of 1/2 for the most recent observation, 1/3 for the second most recent, and 1/6 for third most recent, compute a 3-week weighted moving average for the time series.
   b. Compute the MSE for the weighted moving average in part (a). Do you prefer this weighted moving average to the unweighted moving average? Remember that the MSE for the unweighted moving average is 10.22.
   c. Suppose you are allowed to choose any weights as long as they sum to one. Could you always find a set of weights that would make the MSE smaller for a weighted moving average than for an unweighted moving average? Why or why not?

4. With the gasoline time series data from Table 18.1, show the exponential smoothing forecasts using $\alpha = .1$. Applying the MSE criterion, would you prefer a smoothing constant of $\alpha = .1$ or $\alpha = .2$ for the gasoline sales time series?

5. With a smoothing constant of $\alpha = .2$, equation (18.2) shows that the forecast for the 13th week of the gasoline sales data from Table 18.1 is given by $F_{13} = .2Y_{12} + .8F_{12}$. However, the forecast for week 12 is given by $F_{12} = .2Y_{11} + .8F_{11}$. Thus, we could combine these two results to show that the forecast for the 13th week can be written

$$F_{13} = .2Y_{12} + .8(.2Y_{11} + .8F_{11}) = .2Y_{12} + .16Y_{11} + .64F_{11}$$

   a. Making use of the fact that $F_{11} = .2Y_{10} = .8F_{10}$ (and similarly for F_{10} and F_9), continue to expand the expression for F_{13} until it is written in terms of the past data values $Y_{12}, Y_{11}, Y_{10}, Y_9, Y_8$, and the forecast for period 8.
   b. Refer to the coefficients or weights for the past values $Y_{12}, Y_{11}, Y_{10}, Y_9, Y_8$; what observation can you make about how exponential smoothing weights past data values in arriving at new forecasts? Compare this weighting pattern with the weighting pattern of the moving averages method.

Applications

6. For the Hawkins Company, the monthly percentages of all shipments that were received on time over the past 12 months are 80, 82, 84, 83, 83, 84, 85, 84, 82, 83, 84, and 83.

   a. Compare a 3-month moving average forecast with an exponential smoothing forecast for $\alpha = .2$. Which provides the better forecasts?

   b. What is the forecast for next month?

7. Corporate triple A bond interest rates for 12 consecutive months follow.

9.5   9.3   9.4   9.6   9.8   9.7   9.8   10.5   9.9   9.7   9.6   9.6

   a. Develop 3-month and 4-month moving averages for this time series. Does the 3-month or 4-month moving average provide the better forecasts? Explain.

   b. What is the moving average forecast for the next month?

8. The values of Alabama building contracts ($ millions) for a 12-month period follow.

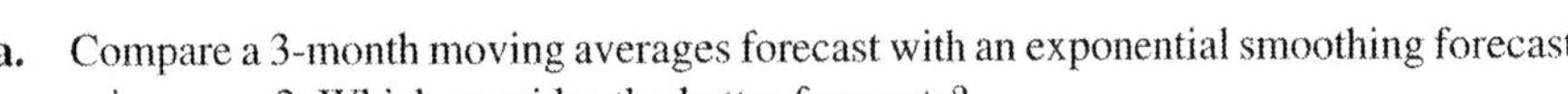

240   350   230   260   280   320   220   310   240   310   240   230

   a. Compare a 3-month moving averages forecast with an exponential smoothing forecast using $\alpha = .2$. Which provides the better forecasts?

   b. What is the forecast for the next month?

9. The following time series shows the sales of a particular product over the past 12 months.

Month	Sales	Month	Sales
1	105	7	145
2	135	8	140
3	120	9	100
4	105	10	80
5	90	11	100
6	120	12	110

   a. Use $\alpha = .3$ to compute the exponential smoothing values for the time series.

   b. Use a smoothing constant of .5 to compute the exponential smoothing values. Does a smoothing constant of .3 or .5 appear to provide the better forecasts?

10. Ten weeks of data on the Commodity Futures Index are 7.35, 7.40, 7.55, 7.56, 7.60, 7.52, 7.52, 7.70, 7.62, and 7.55.

   a. Compute the exponential smoothing values for $\alpha = .2$.

   b. Compute the exponential smoothing values for $\alpha = .3$.

   c. Which exponential smoothing model provides the better forecasts? Forecast week 11.

11. The following data represent 15 quarters of manufacturing capacity utilization (in percentages).

Quarter/Year	Utilization (%)	Quarter/Year	Utilization (%)
1/1998	82.5	1/2000	78.8
2/1998	81.3	2/2000	78.7
3/1998	81.3	3/2000	78.4
4/1998	79.0	4/2000	80.0
1/1999	76.6	1/2001	80.7
2/1999	78.0	2/2001	80.7
3/1999	78.4	3/2001	80.8
4/1999	78.0		

   a. Compute 3- and 4-quarter moving averages for this time series. Which moving average provides the better forecast for the fourth quarter of 2001?

   b. Use smoothing constants of $\alpha = .4$ and $\alpha = .5$ to develop forecasts for the fourth quarter of 2001. Which smoothing constant provides the better forecast?

   c. On the basis of the analyses in parts (a) and (b), which method—moving averages or exponential smoothing—provides the better forecast? Explain.

18.3 TREND PROJECTION

In this section we show how to forecast a time series that has a long-term linear trend. The type of time series for which the trend projection method is applicable shows a consistent increase or decrease over time; it is not stable so the smoothing methods described in the preceding section are not applicable.

Consider the time series for bicycle sales of a particular manufacturer over the past 10 years, as shown in Table 18.6 and Figure 18.8. Note that 21,600 bicycles were sold in year 1, 22,900 were sold in year 2, and so on. In year 10, the most recent year, 31,400 bicycles were sold. Although Figure 18.8 shows some up and down movement over the past 10 years, the time series seems to have an overall increasing or upward trend.

We do not want the trend component of a time series to follow each and every up and down movement. Rather, the trend component should reflect the gradual shifting—in this case, growth—of the time series values. After we view the time series data in Table 18.6 and the graph in Figure 18.8, we might agree that a linear trend as shown in Figure 18.9 provides a reasonable description of the long-run movement in the series.

We use the bicycle sales data to illustrate the calculations involved in applying regression analysis to identify a linear trend. Recall that in the discussion of simple linear regression in Chapter 14, we described how the least squares method is used to find the best straight-line relationship between two variables. That is the methodology we will use to develop the trend line for the bicycle sales time series. Specifically, we will be using regression analysis to estimate the relationship between time and sales volume.

In Chapter 14 the estimated regression equation describing a straight-line relationship between an independent variable x and a dependent variable y was written

$$\hat{y} = b_0 + b_1 x \qquad \text{(18.4)}$$

To emphasize the fact that, in forecasting, the independent variable is time, we will use t in equation (18.4) instead of x; in addition, we will use T_t in place of $\hat{y}$. Thus, for a

TABLE 18.6

BICYCLE SALES
TIME SERIES

Year (t)	Sales (1000s) (Y_t)
1	21.6
2	22.9
3	25.5
4	21.9
5	23.9
6	27.5
7	31.5
8	29.7
9	28.6
10	31.4

FIGURE 18.8   BICYCLE SALES TIME SERIES

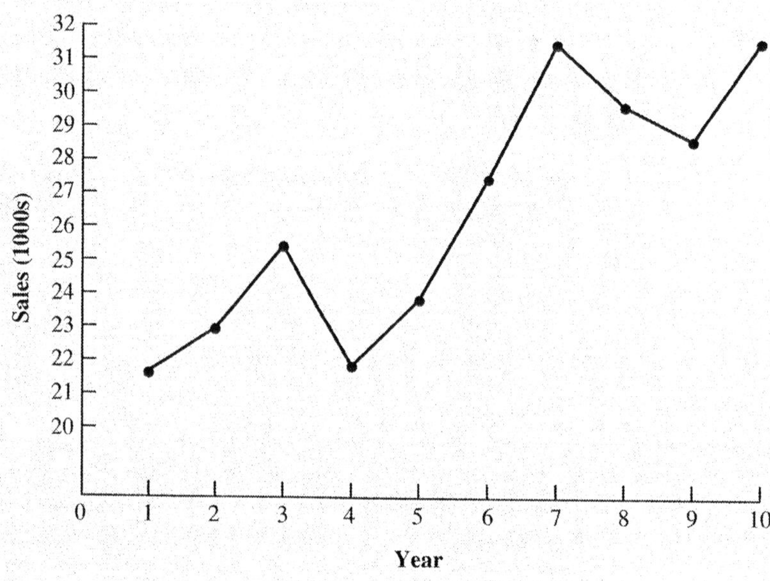

FIGURE 18.9   TREND REPRESENTED BY A LINEAR FUNCTION FOR BICYCLE SALES

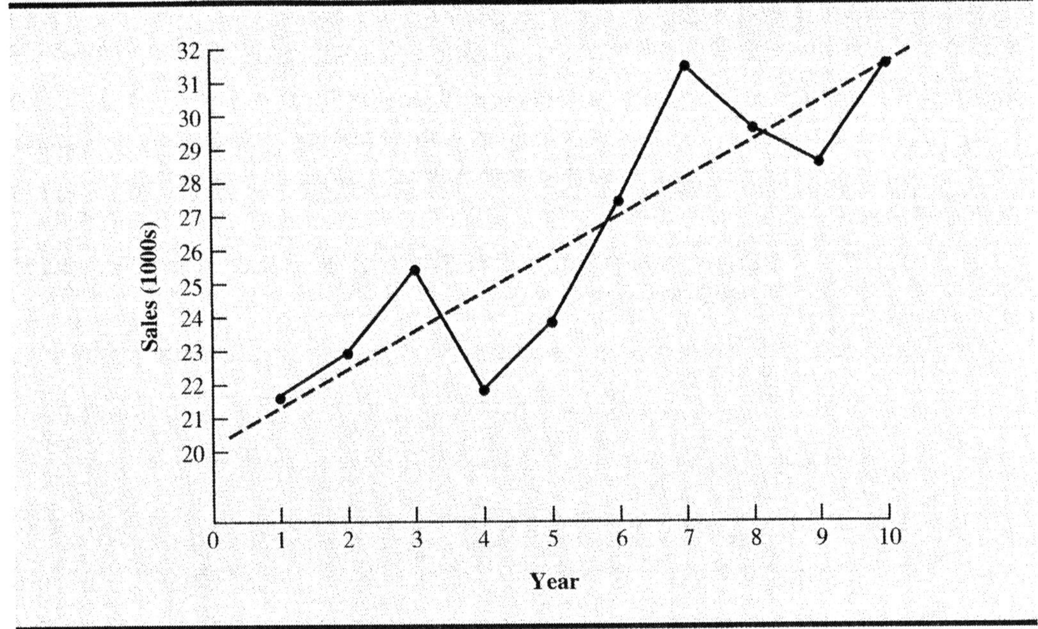

linear trend, the estimated sales volume expressed as a function of time can be written as follows.

Equation for Linear Trend

$$T_t = b_0 + b_1 t \qquad \textbf{(18.5)}$$

where

T_t = trend value of the time series in period t

b_0 = intercept of the trend line

b_1 = slope of the trend line

t = time

In equation (18.5), we will let $t = 1$ for the time of the first observation on the time series data, $t = 2$ for the time of the second observation, and so on. Note that for the time series on bicycle sales, $t = 1$ corresponds to the oldest time series value and $t = 10$ corresponds to the most recent year's data. Formulas for computing the estimated regression coefficients (b_1 and b_0) in (18.5) are shown below.

Computing the Slope (b_1) and Intercept (b_0)

$$b_1 = \frac{\Sigma t Y_t - (\Sigma t \Sigma Y_t)/n}{\Sigma t^2 - (\Sigma t)^2/n} \qquad \textbf{(18.6)}$$

$$b_0 = \bar{Y} - b_1 \bar{t} \qquad \textbf{(18.7)}$$

where

$$Y_t = \text{value of the time series in period } t$$
$$n = \text{number of periods}$$
$$\bar{Y} = \text{average value of the time series; that is, } \bar{Y} = \Sigma Y_t/n$$
$$\bar{t} = \text{average value of } t; \text{ that is, } \bar{t} = \Sigma t/n$$

Using equations (18.6) and (18.7) and the bicycle sales data of Table 18.6, we can compute b_0 and b_1 as follows:

t	Y_t	tY_t	t^2
1	21.6	21.6	1
2	22.9	45.8	4
3	25.5	76.5	9
4	21.9	87.6	16
5	23.9	119.5	25
6	27.5	165.0	36
7	31.5	220.5	49
8	29.7	237.6	64
9	28.6	257.4	81
10	31.4	314.0	100
Totals 55	264.5	1545.5	385

$$\bar{t} = \frac{55}{10} = 5.5$$

$$\bar{Y} = \frac{264.5}{10} = 26.45$$

$$b_1 = \frac{1545.5 - (55)(264.5)/10}{385 - (55)^2/10} = 1.10$$

$$b_0 = 26.45 - 1.10(5.5) = 20.4$$

Therefore,

$$T_t = 20.4 + 1.1t \tag{18.8}$$

is the expression for the linear trend component for the bicycle sales time series.

Before the trend equation is used to develop a forecast, a statistical test of significance (see Chapter 14) should be conducted. In practice, such a test would be a routine part of fitting the trend line.

The slope of 1.1 indicates that over the past 10 years the firm has had an average growth in sales of about 1100 units per year. If we assume that the past 10-year trend in sales is a good indicator of the future, (18.8) can be used to project the trend component of the time series. For example, substituting $t = 11$ into (18.8) yields next year's trend projection, T_{11}.

$$T_{11} = 20.4 + 1.1(11) = 32.5$$

Thus, using the trend component only, we would forecast sales of 32,500 bicycles next year.

The use of a linear function to model the trend is common. However, as we discussed previously, sometimes time series have a curvilinear, or nonlinear, trend similar to those in

FIGURE 18.10   SOME POSSIBLE FORMS OF NONLINEAR TREND PATTERNS

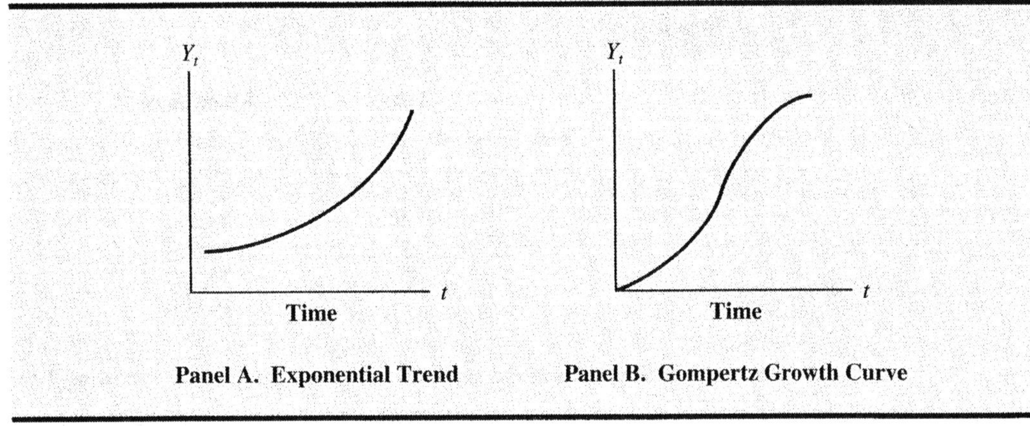

Figure 18.10. In Chapter 16 we discussed how regression analysis can be used to model curvilinear relationships of the type shown in panel A of Figure 18.10. More advanced texts discuss in detail how to develop regression models for more complex relationships, such as the one shown in panel B of Figure 18.10.

EXERCISES

Methods

12. Consider the following time series.

t	1	2	3	4	5
Y_t	6	11	9	14	15

Develop an equation for the linear trend component of this time series. What is the forecast for $t = 6$?

13. Consider the following time series.

t	1	2	3	4	5	6
Y_t	205	202	195	190	191	188

Develop an equation for the linear trend component for this time series. What is the forecast for $t = 7$?

Applications

14. The enrollment data (1000s) for a state college over the past 6 years are shown.

Year	1	2	3	4	5	6
Enrollment	20.5	20.2	19.5	19.0	19.1	18.8

Develop the equation for the linear trend component of this time series. Comment on what is happening to enrollment at this institution.

15. The following table gives average attendance figures for home football games at a major university for the past 7 years. Develop the equation for the linear trend component of this time series.

Year	Attendance
1	28,000
2	30,000
3	31,500
4	30,400
5	30,500
6	32,200
7	30,800

16. Automobile sales at B.J. Scott Motors, Inc., provided the following 10-year time series.

Year	Sales
1	400
2	390
3	320
4	340
5	270
6	260
7	300
8	320
9	340
10	370

Plot the time series and comment on the appropriateness of a linear trend. What type of functional form do you believe would be most appropriate for the trend pattern of this time series?

17. The president of a small manufacturing firm has been concerned about the continual increase in manufacturing costs over the past several years. The following figures provide a time series of the cost per unit for the firm's leading product over the past 8 years.

Year	Cost/Unit ($)
1	20.00
2	24.50
3	28.20
4	27.50
5	26.60
6	30.00
7	31.00
8	36.00

a. Show a graph of this time series. Does a linear trend appear to be present?
b. Develop the equation for the linear trend component of the time series. What is the average cost increase that the firm has been realizing per year?

18. Earnings per share for the Walgreen Company for a 10-year period follow.

.64 .73 .94 1.14 1.33 1.53 1.67 1.68 2.10 2.50

a. Use a linear trend projection to forecast this time series for the coming year.
b. What does this time series analysis tell you about the Walgreen Company? Do the historical data indicate that the Walgreen Company is a good investment?

19. In the late 1990s many firms began downsizing in order to reduce their costs. One of the results of these cost-cutting measures has been a decline in the percentage of private-industry jobs that are managerial. The following data show the percentage of females who are managers from 1990 to 1995 (*The Wall Street Journal Almanac*, 1998).

Year	1990	1991	1992	1993	1994	1995
Percentage	7.45	7.53	7.52	7.65	7.62	7.73

a. Develop a linear trend equation for this time series.
b. Use the trend equation to estimate the percentage of females who are managers for 1996 and 1997.

20. Gross revenue data ($ millions) for Regional Airlines for a 10-year period follow.

Year	Revenue	Year	Revenue
1	2428	6	4264
2	2951	7	4738
3	3533	8	4460
4	3618	9	5318
5	3616	10	6915

a. Develop a linear trend equation for this time series. Comment on what the equation tells about the gross revenue for Regional Airlines for the 10-year period.
b. Provide the forecasts for gross revenue for years 11 and 12.

21. ACT Networks, Inc., develops, markets, manufactures, and sells integrated wide-area network access products. The following are annual sales data from 1992 to 1997 (*Stock Investor Pro*, American Association of Individual Investors, August 31, 1997).

Year	Sales ($millions)
1992	5.4
1993	6.2
1994	12.7
1995	20.6
1996	28.4
1997	44.9

a. Develop a linear trend equation for this time series.
b. What is the firm's average increase in sales per year?
c. Use the trend equation to forecast sales for 1998.

18.4 TREND AND SEASONAL COMPONENTS

We have shown how to forecast a time series that has a trend component. In this section we extend the discussion by showing how to forecast a time series that has both trend and seasonal components.

Many situations in business and economics involve period-to-period comparisons. For instance, we might be interested to learn that unemployment is up 2% compared to last month, steel production is up 5% over last month, or that the production of electric power is down 3% from the previous month. Care must be exercised in using such information, however, because whenever a seasonal influence is present, such comparisons may be misleading. For instance, the fact that electric power consumption is down 3% from August to September might be only the seasonal effect associated with a decrease in the use of air conditioning and not because of a long-term decline in the use of electric power. Indeed, after adjusting for the seasonal effect, we might even find that the use of electric power has increased.

Removing the seasonal effect from a time series is known as deseasonalizing the time series. After we do so, period-to-period comparisons are more meaningful and can help identify whether a trend exists. The approach we take in this section is appropriate in situations when only seasonal effects are present or in situations when both seasonal and trend components are present. The first step is to compute seasonal indexes and use them to deseasonalize the data. Then, if a trend is apparent in the deseasonalized data, we use regression analysis on the deseasonalized data to estimate the trend component.

Multiplicative Model

In addition to a trend component (T) and a seasonal component (S), we will assume that the time series has an irregular component (I). The irregular component accounts for any random effects in the time series that cannot be explained by the trend and seasonal components. Using T_t, S_t, and I_t to identify the trend, seasonal, and irregular components at time t, we will assume that the time series value, denoted Y_t, can be described by the following **multiplicative time series model.**

$$Y_t = T_t \times S_t \times I_t \qquad (18.9)$$

In this model, T_t is the trend measured in units of the item being forecast. However, the S_t and I_t components are measured in relative terms, with values above 1.00 indicating effects above the trend and values below 1.00 indicating effects below the trend.

We will illustrate the use of the multiplicative model with trend, seasonal, and irregular components by working with the quarterly data in Table 18.7 and Figure 18.11. These data show television set sales (in thousands of units) for a particular manufacturer over the past 4 years. We begin by showing how to identify the seasonal component of the time series.

Calculating the Seasonal Indexes

Figure 18.11 indicates that sales are lowest in the second quarter of each year and increase in quarters 3 and 4. Thus, we conclude that a seasonal pattern exists for television set sales. We can begin the computational procedure used to identify each quarter's seasonal influence by computing a moving average to separate the combined seasonal and irregular components, S_t and I_t, from the trend component T_t.

To do so, we use one year of data in each calculation. Because we are working with a quarterly series, we will use four data values in each moving average. The moving average calculation for the first four quarters of the television set sales data is

$$\text{First Moving Average} = \frac{4.8 + 4.1 + 6.0 + 6.5}{4} = \frac{21.4}{4} = 5.35$$

Note that the moving average calculation for the first four quarters yields the average quarterly sales over year 1 of the time series. Continuing the moving average calculation, we next add the 5.8 value for the first quarter of year 2 and drop the 4.8 for the first quarter of year 1. Thus, the second moving average is

$$\text{Second Moving Average} = \frac{4.1 + 6.0 + 6.5 + 5.8}{4} = \frac{22.4}{4} = 5.60$$

Similarly, the third moving average calculation is $(6.0 + 6.5 + 5.8 + 5.2)/4 = 5.875$.

Before we proceed with the moving average calculations for the entire time series, we return to the first moving average calculation, which resulted in a value of 5.35. The 5.35 value represents an average quarterly sales volume (across all seasons) for year 1. As we

TABLE 18.7

QUARTERLY DATA FOR TELEVISION SET SALES

Year	Quarter	Sales (1000s)
1	1	4.8
	2	4.1
	3	6.0
	4	6.5
2	1	5.8
	2	5.2
	3	6.8
	4	7.4
3	1	6.0
	2	5.6
	3	7.5
	4	7.8
4	1	6.3
	2	5.9
	3	8.0
	4	8.4

FIGURE 18.11   QUARTERLY TELEVISION SET SALES TIME SERIES

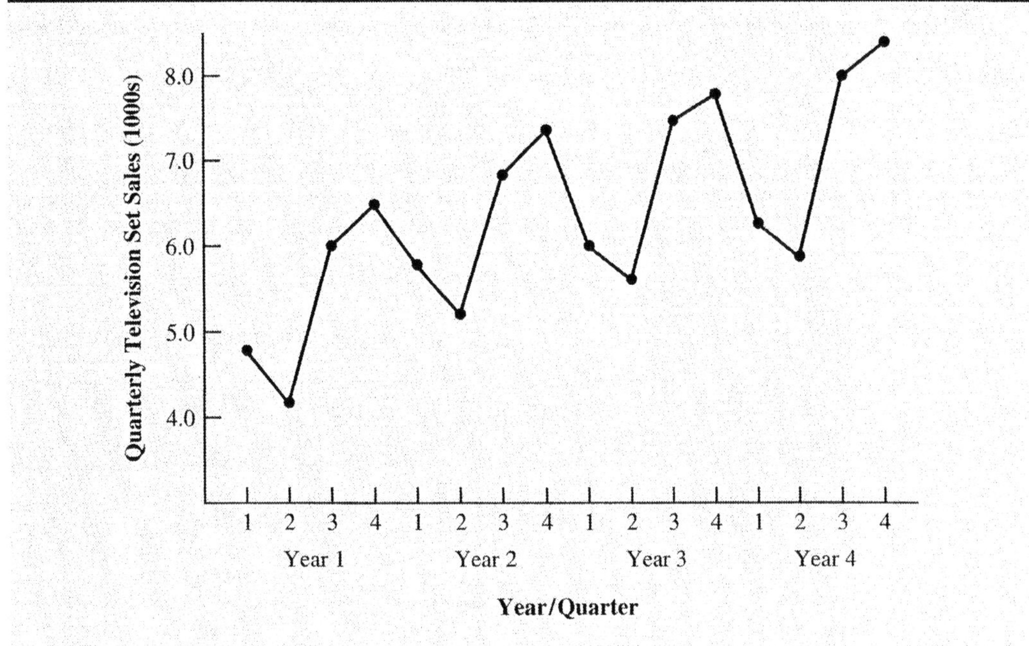

look back at the calculation of the 5.35 value, associating 5.35 with the "middle" quarter of the moving average group makes sense. Note, however, that we encounter some difficulty in identifying the middle quarter; with four quarters in the moving average there is no middle quarter. The 5.35 value corresponds to the last half of quarter 2 and the first half of quarter 3. Similarly, if we go to the next moving average value of 5.60, the middle corresponds to the last half of quarter 3 and the first half of quarter 4.

Recall that the reason for computing moving averages is to isolate the combined seasonal and irregular components. However, the moving average values we have computed do not correspond directly to the original quarters of the time series. We can resolve this difficulty by using the midpoints between successive moving average values. For example, if 5.35 corresponds to the first half of quarter 3 and 5.60 corresponds to the last half of quarter 3, we can use $(5.35 + 5.60)/2 = 5.475$ as the moving average value for quarter 3. Similarly, we associate a moving average value of $(5.60 + 5.875)/2 = 5.738$ with quarter 4. The result is a *centered moving average.* Table 18.8 shows a complete summary of the moving average calculations for the television set sales data.

If the number of data points in a moving average calculation is an odd number, the middle point will correspond to one of the periods in the time series. In such cases, we would not have to center the moving average values to correspond to a particular time period as we have done in the calculations in Table 18.8.

What do the centered moving averages in Table 18.8 tell us about this time series? Figure 18.12 is a plot of the actual time series values and the centered moving average values. Note particularly how the centered moving average values tend to "smooth out" both the seasonal and irregular fluctuations in the time series. The moving average values computed for 4 quarters of data do not include the fluctuations due to seasonal influences because the seasonal effect has been averaged out. Each point in the centered moving average represents the value of the time series as though there were no seasonal or irregular influence.

By dividing each time series observation by the corresponding centered moving average, we can identify the seasonal irregular effect in the time series. For example, the third quarter

TABLE 18.8   CENTERED MOVING AVERAGE CALCULATIONS FOR THE TELEVISION
SET SALES TIME SERIES

Year	Quarter	Sales (1000s)	4-Quarter Moving Average	Centered Moving Average
1	1	4.8		
	2	4.1		
			5.350	
	3	6.0		5.475
			5.600	
	4	6.5		5.738
			5.875	
2	1	5.8		5.975
			6.075	
	2	5.2		6.188
			6.300	
	3	6.8		6.325
			6.350	
	4	7.4		6.400
			6.450	
3	1	6.0		6.538
			6.625	
	2	5.6		6.675
			6.725	
	3	7.5		6.763
			6.800	
	4	7.8		6.838
			6.875	
4	1	6.3		6.938
			7.000	
	2	5.9		7.075
			7.150	
	3	8.0		
	4	8.4		

of year 1 shows 6.0/5.475 = 1.096 as the combined seasonal irregular value. Table 18.9 summarizes the seasonal irregular values for the entire time series.

Consider the third quarter. The results from years 1, 2, and 3 show third-quarter values of 1.096, 1.075, and 1.109, respectively. Thus, in all cases, the seasonal irregular value appears to have an above-average influence in the third quarter. With the year-to-year fluctuations in the seasonal irregular value attributable primarily to the irregular component, we can average the computed values to eliminate the irregular influence and obtain an estimate of the third-quarter seasonal influence.

$$\text{Seasonal Effect of Third Quarter} = \frac{1.096 + 1.075 + 1.109}{3} = 1.09$$

We refer to 1.09 as the *seasonal index* for the third quarter. In Table 18.10 we summarize the calculations involved in computing the seasonal indexes for the television set sales time

FIGURE 18.12   QUARTERLY TELEVISION SET SALES TIME SERIES AND CENTERED MOVING AVERAGE

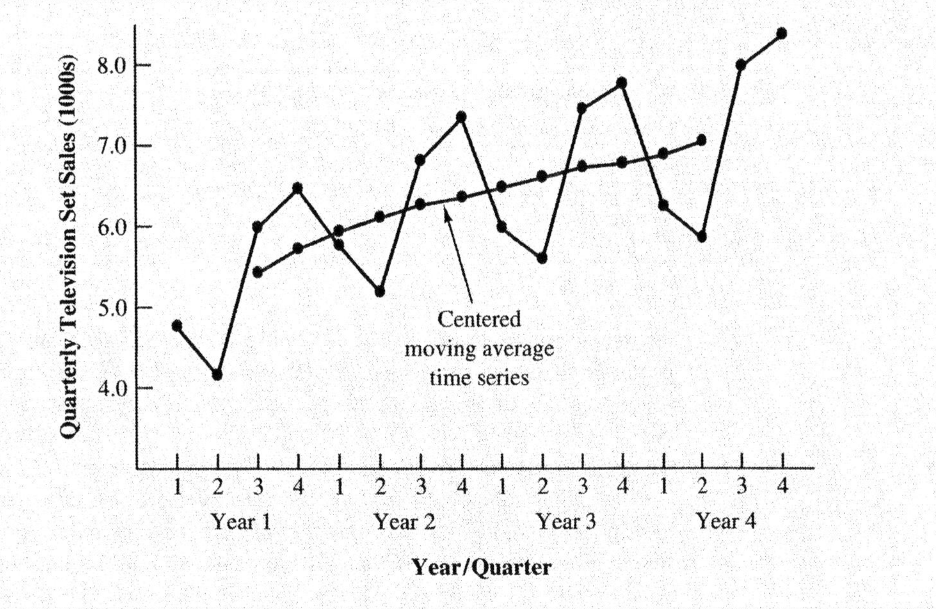

series. Thus, the seasonal indexes for the four quarters are: quarter 1, .93; quarter 2, .84; quarter 3, 1.09; and quarter 4, 1.14.

Interpretation of the values in Table 18.10 provides some observations about the seasonal component in television set sales. The best sales quarter is the fourth quarter, with sales averaging 14% above the average quarterly value. The worst, or slowest, sales quarter is the second quarter; its seasonal index of .84 shows that the sales average is 16% below

TABLE 18.9   SEASONAL IRREGULAR VALUES FOR THE TELEVISION SET SALES TIME SERIES

Year	Quarter	Sales (1000s)	Centered Moving Average	Seasonal Irregular Value
1	1	4.8		
	2	4.1		
	3	6.0	5.475	1.096
	4	6.5	5.738	1.133
2	1	5.8	5.975	.971
	2	5.2	6.188	.840
	3	6.8	6.325	1.075
	4	7.4	6.400	1.156
3	1	6.0	6.538	.918
	2	5.6	6.675	.839
	3	7.5	6.763	1.109
	4	7.8	6.838	1.141
4	1	6.3	6.938	.908
	2	5.9	7.075	.834
	3	8.0		
	4	8.4		

TABLE 18.10   SEASONAL INDEX CALCULATIONS FOR THE TELEVISION SET SALES
TIME SERIES

Quarter	Seasonal Irregular Component Values $(S_t I_t)$	Seasonal Index (S_t)
1	.971, .918, .908	.93
2	.840, .839, .834	.84
3	1.096, 1.075, 1.109	1.09
4	1.133, 1.156, 1.141	1.14

the average quarterly sales. The seasonal component corresponds clearly to the intuitive expectation that television viewing interest and thus television purchase patterns tend to peak in the fourth quarter because of the coming winter season and reduction in outdoor activities. The low second-quarter sales reflect the reduced interest in television viewing due to the spring and presummer activities of potential customers.

One final adjustment is sometimes necessary in obtaining the seasonal indexes. The multiplicative model requires that the average seasonal index equal 1.00, so the sum of the four seasonal indexes in Table 18.10 must equal 4.00. In other words the seasonal effects must even out over the year. The average of the seasonal indexes in our example is equal to 1.00, and hence this type of adjustment is not necessary. In other cases, a slight adjustment may be necessary. To make the adjustment, multiply each seasonal index by the number of seasons divided by the sum of the unadjusted seasonal indexes. For instance, for quarterly data multiply each seasonal index by 4/(sum of the unadjusted seasonal indexes). Some of the exercises will require this adjustment to obtain the appropriate seasonal indexes.

Deseasonalizing the Time Series

The purpose of finding seasonal indexes is to remove the seasonal effects from a time series. This process is referred to as *deseasonalizing* the time series. Economic time series adjusted for seasonal variations (**deseasonalized time series**) are often reported in publications such as the *Survey of Current Business*, *The Wall Street Journal*, and *Business Week*. Using the notation of the multiplicative model, we have

With deseasonalized data, comparing sales in successive periods makes sense. With data that have not been deseasonalized, relevant comparisons can often be made between sales in the current period and sales in the same period one year ago.

$$Y_t = T_t \times S_t \times I_t$$

By dividing each time series observation by the corresponding seasonal index, we have removed the effect of season from the time series. The deseasonalized time series for television set sales is summarized in Table 18.11. A graph of the deseasonalized television set sales time series is shown in Figure 18.13.

Using the Deseasonalized Time Series to Identify Trend

Although the graph in Figure 18.13 shows some random up and down movement over the past 16 quarters, the time series seems to have an upward linear trend. To identify this trend, we will use the same procedure as in the preceding section; in this case, the data are quarterly deseasonalized sales values. Thus, for a linear trend, the estimated sales volume expressed as a function of time is

$$T_t = b_0 + b_1 t$$

FIGURE 18.13 DESEASONALIZED TELEVISION SET SALES TIME SERIES

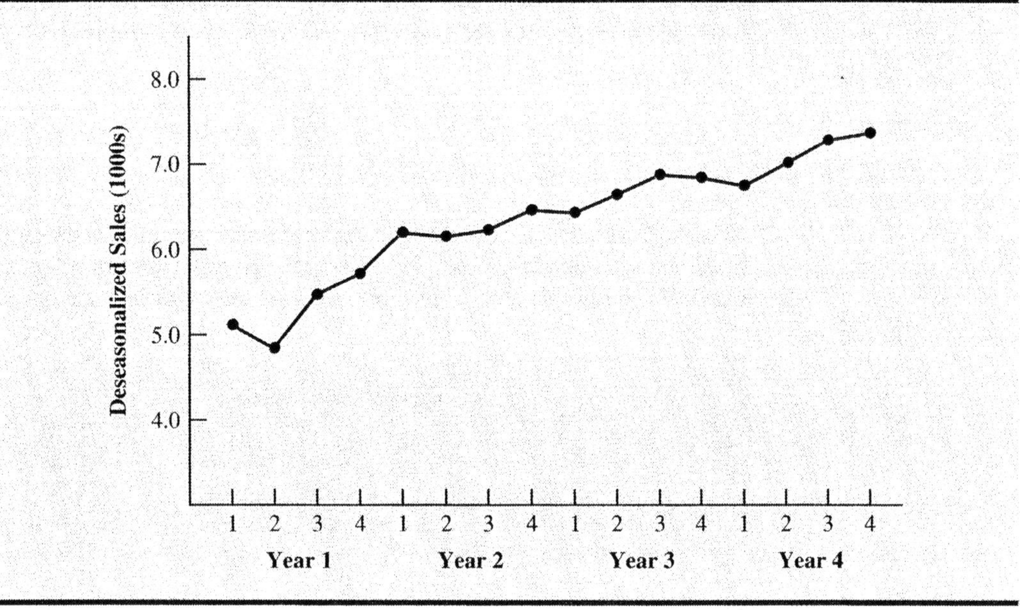

where

T_t = trend value for television set sales in period t

b_0 = intercept of the trend line

b_1 = slope of the trend line

As before, $t = 1$ corresponds to the time of the first observation for the time series, $t = 2$ corresponds to the time of the second observation, and so on. Thus, for the deseasonalized

TABLE 18.11 DESEASONALIZED VALUES FOR THE TELEVISION SET SALES TIME SERIES

Year	Quarter	Sales (1000s) (Y_t)	Seasonal Index (S_t)	Deseasonalized Sales $(Y_t/S_t = T_t I_t)$
1	1	4.8	.93	5.16
	2	4.1	.84	4.88
	3	6.0	1.09	5.50
	4	6.5	1.14	5.70
2	1	5.8	.93	6.24
	2	5.2	.84	6.19
	3	6.8	1.09	6.24
	4	7.4	1.14	6.49
3	1	6.0	.93	6.45
	2	5.6	.84	6.67
	3	7.5	1.09	6.88
	4	7.8	1.14	6.84
4	1	6.3	.93	6.77
	2	5.9	.84	7.02
	3	8.0	1.09	7.34
	4	8.4	1.14	7.37

television set sales time series, $t = 1$ corresponds to the first deseasonalized quarterly sales value and $t = 16$ corresponds to the most recent deseasonalized quarterly sales value. The formulas for computing the value of b_0 and the value of b_1 follow.

$$b_1 = \frac{\Sigma t Y_t - (\Sigma t \Sigma Y_t)/n}{\Sigma t^2 - (\Sigma t)^2/n}$$

$$b_0 = \bar{Y} - b_1 \bar{t}$$

Note, however, that Y_t now refers to the deseasonalized time series value at time t and not to the actual value of the time series. Using the given relationships for b_0 and b_1 and the deseasonalized sales data of Table 18.11, we have the following calculations.

t	Y_t (Deseasonalized)	tY_t	t^2
1	5.16	5.16	1
2	4.88	9.76	4
3	5.50	16.50	9
4	5.70	22.80	16
5	6.24	31.20	25
6	6.19	37.14	36
7	6.24	43.68	49
8	6.49	51.92	64
9	6.45	58.05	81
10	6.67	66.70	100
11	6.88	75.68	121
12	6.84	82.08	144
13	6.77	88.01	169
14	7.02	98.28	196
15	7.34	110.10	225
16	7.37	117.92	256
Totals 136	101.74	914.98	1496

where

$$\bar{t} = \frac{136}{16} = 8.5$$

$$\bar{Y} = \frac{101.74}{16} = 6.359$$

$$b_1 = \frac{914.98 - (136)(101.74)/16}{1496 - (136)^2/16} = 0.148$$

$$b_0 = 6.359 - 0.148(8.5) = 5.101$$

Therefore,

$$T_t = 5.101 + 0.148t$$

is the expression for the linear trend component of the time series.

The slope of 0.148 indicates that over the past 16 quarters, the firm has had an average deseasonalized growth in sales of around 148 sets per quarter. If we assume that the past 16-quarter trend in sales data is a reasonably good indicator of the future, this equation can be used to project the trend component of the time series for future quarters. For example, substituting $t = 17$ into the equation yields next quarter's trend projection, T_{17}.

$$T_{17} = 5.101 + 0.148(17) = 7.617$$

Thus, the trend component yields a sales forecast of 7617 television sets for the next quarter. Similarly, the trend component produces sales forecasts of 7765, 7913, and 8061 television sets in quarters 18, 19, and 20, respectively.

Seasonal Adjustments

The final step in developing the forecast when both trend and seasonal components are present is to use the seasonal index to adjust the trend projection. Returning to the television set sales example, we have a trend projection for the next four quarters. Now we must adjust the forecast for the seasonal effect. The seasonal index for the first quarter of year 5 ($t = 17$) is 0.93, so we obtain the quarterly forecast by multiplying the forecast based on trend ($T_{17} = 7617$) by the seasonal index (0.93). Thus, the forecast for the next quarter is $7617(0.93) = 7084$. Table 18.12 gives the quarterly forecast for quarters 17 through 20. The high-volume fourth quarter has a 9190-unit forecast, and the low-volume second quarter has a 6523-unit forecast.

Models Based on Monthly Data

In the preceding television set sales example, we used quarterly data to illustrate the computation of seasonal indexes. However, many businesses use monthly rather than quarterly forecasts. In such cases, the procedures introduced in this section can be applied with minor modifications. First, a 12-month moving average replaces the 4-quarter moving average; second, 12 monthly seasonal indexes, rather than four quarterly seasonal indexes, must be computed. Other than these changes, the computational and forecasting procedures are identical.

Cyclical Component

Mathematically, the multiplicative model of (18.9) can be expanded to include a cyclical component.

$$Y_t = T_t \times C_t \times S_t \times I_t \tag{18.10}$$

The cyclical component, like the seasonal component, is expressed as a percentage of trend. As mentioned in Section 18.1, this component is attributable to multiyear cycles in the time series. It is analogous to the seasonal component, but over a longer period of time. However, because of the length of time involved, obtaining enough relevant data to estimate the cyclical component is often difficult. Another difficulty is that cycles usually vary in length. We leave further discussion of the cyclical component to texts on forecasting methods.

TABLE 18.12   QUARTERLY FORECASTS FOR THE TELEVISION SET SALES TIME SERIES

Year	Quarter	Trend Forecast	Seasonal Index (see Table 18.11)	Quarterly Forecast
5	1	7617	.93	$(7617)(.93) = 7084$
	2	7765	.84	$(7765)(.84) = 6523$
	3	7913	1.09	$(7913)(1.09) = 8625$
	4	8061	1.14	$(8061)(1.14) = 9190$

EXERCISES

Methods

22. Consider the following time series data.

Quarter	Year 1	Year 2	Year 3
1	4	6	7
2	2	3	6
3	3	5	6
4	5	7	8

    a. Show the 4-quarter and centered moving average values for this time series.
    b. Compute seasonal indexes for the 4 quarters.

Applications

23. The quarterly sales data (number of copies sold) for a college textbook over the past 3 years follow.

Quarter	Year 1	Year 2	Year 3
1	1690	1800	1850
2	940	900	1100
3	2625	2900	2930
4	2500	2360	2615

    a. Show the 4-quarter and centered moving average values for this time series.
    b. Compute seasonal indexes for the 4 quarters.
    c. When does the textbook publisher have the largest seasonal index? Does this appear reasonable? Explain.

24. Identify the monthly seasonal indexes for the 3 years of expenses for a six-unit apartment house in southern Florida as given here. Use a 12-month moving average calculation.

	Expenses		
	Year 1	Year 2	Year 3
January	170	180	195
February	180	205	210
March	205	215	230
April	230	245	280
May	240	265	290
June	315	330	390
July	360	400	420
August	290	335	330
September	240	260	290
October	240	270	295
November	230	255	280
December	195	220	250

25. Air pollution control specialists in southern California monitor the amount of ozone, carbon dioxide, and nitrogen dioxide in the air on an hourly basis. The hourly time series data exhibit seasonality, with the levels of pollutants showing patterns over the hours in the day. On July 15, 16, and 17, the following levels of nitrogen dioxide were observed in the downtown area for the 12 hours from 6:00 A.M. to 6:00 P.M.

July 15:	25	28	35	50	60	60	40	35	30	25	25	20
July 16:	28	30	35	48	60	65	50	40	35	25	20	20
July 17:	35	42	45	70	72	75	60	45	40	25	25	25

   a. Identify the hourly seasonal indexes for the 12 readings each day.

   b. With the seasonal indexes from part (a), the data were deseasonalized; the trend equation developed for the deseasonalized data was $T_t = 32.983 + .3922t$. Using the trend component only, develop forecasts for the 12 hours for July 18.

   c. Use the seasonal indexes from part (a) to adjust the trend forecasts developed in part (b).

26. Electric power consumption is measured in kilowatt-hours (kWh). The local utility company has an interrupt program whereby commercial customers that participate receive favorable rates but must agree to cut back consumption if the utility requests them to do so. Timko Products cut back consumption at 12:00 noon Thursday. To assess the savings, the utility must estimate Timko's usage without the interrupt. The period of interrupted service was from noon to 8:00 P.M. Data on electric power consumption for the previous 72 hours are available.

Time Period	Monday	Tuesday	Wednesday	Thursday
12–4 A.M.	—	19,281	31,209	27,330
4–8 A.M.	—	33,195	37,014	32,715
8–12 noon	—	99,516	119,968	152,465
12–4 P.M.	124,299	123,666	156,033	
4–8 P.M.	113,545	111,717	128,889	
8–12 midnight	41,300	48,112	73,923	

   a. Is there a seasonal effect over the 24-hour period? Compute seasonal indexes for the six 4-hour periods.

   b. Use trend adjusted for seasonal indexes to estimate Timko's normal usage over the period of interrupted service.

18.5 REGRESSION ANALYSIS

In the discussion of regression analysis in Chapters 14, 15, and 16, we showed how one or more independent variables could be used to predict the value of a single dependent variable. Looking at regression analysis as a forecasting tool, we can view the time series value that we want to forecast as the dependent variable. Hence, if we can identify a good set of related independent, or predictor, variables we may be able to develop an estimated regression equation for predicting or forecasting the time series.

The approach we used in Section 18.3 to fit a linear trend line to the bicycle sales time series is a special case of regression analysis. In that example, two variables—bicycle sales and time—were shown to be linearly related.* The inherent complexity of most real-world problems necessitates the consideration of more than one variable to predict the variable of interest. The statistical technique known as multiple regression analysis can be used in such situations.

Recall that to develop an estimated multiple regression equation, we need a sample of observations for the dependent variable and all independent variables. In time series analysis the n periods of time series data provide a sample of n observations on each variable that

---

*In a purely technical sense, the number of bicycles sold is not viewed as being related to time; instead, time is used as a surrogate for variables to which the number of bicycles sold is actually related but which are either unknown or too difficult or too costly to measure.

can be used in the analysis. For a function involving k independent variables, we use the following notation.

$$Y_t = \text{value of the time series in period } t$$
$$x_{1t} = \text{value of independent variable 1 in period } t$$
$$x_{2t} = \text{value of independent variable 2 in period } t$$
$$\vdots$$
$$x_{kt} = \text{value of independent variable } k \text{ in period } t$$

The n periods of data necessary to develop the estimated regression equation would appear as shown in the following table.

Period	Time Series (Y_t)	Value of Independent Variables						
		x_{1t}	x_{2t}	x_{3t}	.	.	.	x_{kt}
1	Y_1	x_{11}	x_{21}	x_{31}	.	.	.	x_{k1}
2	Y_2	x_{12}	x_{22}	x_{32}	.	.	.	x_{k2}
.	.	.	.	.	.	.	.	.
.	.	.	.	.	.	.	.	.
.	.	.	.	.	.	.	.	.
n	Y_n	x_{1n}	x_{2n}	x_{3n}	.	.	.	x_{kn}

As you might imagine, several choices are possible for the independent variables in a forecasting model. One possible choice for an independent variable is simply time. It was the choice made in Section 18.3 when we estimated the trend of the time series using a linear function of the independent variable time. Letting $x_{1t} = t$, we obtain an estimated regression equation of the form

$$\hat{Y}_t = b_0 + b_1 t$$

where $\hat{Y}_t$ is the estimate of the time series value Y_t and where b_0 and b_1 are the estimated regression coefficients. In a more complex model, additional terms could be added corresponding to time raised to other powers. For example, if $x_{2t} = t^2$ and $x_{3t} = t^3$, the estimated regression equation would become

$$\hat{Y}_t = b_0 + b_1 x_{1t} + b_2 x_{2t} + b_3 x_{3t}$$
$$= b_0 + b_1 t + b_2 t^2 + b_3 t^3$$

Note that this model provides a forecast of a time series with curvilinear characteristics over time.

Other regression-based forecasting models have a mixture of economic and demographic independent variables. For example, in forecasting sales of refrigerators, we might select the following independent variables.

$$x_{1t} = \text{price in period } t$$
$$x_{2t} = \text{total industry sales in period } t - 1$$
$$x_{3t} = \text{number of building permits for new houses in period } t - 1$$
$$x_{4t} = \text{population forecast for period } t$$
$$x_{5t} = \text{advertising budget for period } t$$

According to the usual multiple regression procedure, an estimated regression equation with five independent variables would be used to develop forecasts.

Spyros Makridakis, a noted forecasting expert, has conducted research showing that simple techniques usually outperform more complex procedures for short-term forecasting. Using a more sophisticated and expensive procedure will not guarantee better forecasts.

Whether a regression approach provides a good forecast depends largely on how well we are able to identify and obtain data for independent variables that are closely related to the time series. Generally, during the development of an estimated regression equation, we will want to consider many possible sets of independent variables. Thus, part of the regression analysis procedure should be the selection of the set of independent variables that provides the best forecasting model.

In the chapter introduction we stated that the **causal forecasting models** use other time series related to the one being forecast in an effort to explain the cause of a time series' behavior. Regression analysis is the tool most often used in developing such causal models. The related time series become the independent variables, and the time series being forecast is the dependent variable.

In another type of regression-based forecasting model, the independent variables are all previous values of the same time series. For example, if the time series values are denoted $Y_1, Y_2, \ldots, Y_n$, then with a dependent variable Y_t, we might try to find an estimated regression equation relating Y_t to the most recent times series values Y_{t-1}, Y_{t-2}, and so on. With the three most recent periods as independent variables, the estimated regression equation would be

$$\hat{Y}_t = b_0 + b_1 Y_{t-1} + b_2 Y_{t-2} + b_3 Y_{t-3}$$

Regression models in which the independent variables are previous values of the time series are referred to as **autoregressive models.**

Finally, another regression-based forecasting approach incorporates a mixture of the independent variables previously discussed. For example, we might select a combination of time variables, some economic/demographic variables, and some previous values of the time series variable itself.

18.6   QUALITATIVE APPROACHES

If historical data are not available, managers must use a qualitative technique to develop forecasts. But the cost of using qualitative techniques can be high because of the time commitment required from the people involved.

In the preceding sections we discussed several types of quantitative forecasting methods. Most of those techniques require historical data on the variable of interest, so they cannot be applied when historical data are not available. Furthermore, even when such data are available, a significant change in environmental conditions affecting the time series may make the use of past data questionable in predicting future values of the time series. For example, a government-imposed gasoline rationing program would raise questions about the validity of a gasoline sales forecast based on historical data. Qualitative forecasting techniques afford an alternative in these and other cases.

Delphi Method

One of the most commonly used qualitative forecasting techniques is the **Delphi method,** originally developed by a research group at the Rand Corporation. It is an attempt to develop forecasts through "group consensus." In its usual application, the members of a panel of experts—all of whom are physically separated from and unknown to each other—are asked to respond to a series of questionnaires. The responses from the first questionnaire are tabulated and used to prepare a second questionnaire that contains information and opinions of the entire group. Each respondent is then asked to reconsider and possibly revise his or her previous response in light of the group information provided. This process continues until the coordinator feels that some degree of consensus has been reached. The goal of the

Delphi method is not to produce a single answer as output, but instead to produce a relatively narrow spread of opinions within which the majority of experts concur.

Expert Judgment

Empirical evidence and theoretical arguments suggest that between 5 and 20 experts should be used in judgmental forecasting. However, in situations involving exponential growth, judgmental forecasts may be inappropriate.

Qualitative forecasts often are based on the judgment of a single expert or represent the consensus of a group of experts. For example, each year a group of experts at Merrill Lynch gather to forecast the level of the Dow Jones Industrial Average and the prime rate for the next year. In doing so, the experts individually consider information that they believe will influence the stock market and interest rates; then they combine their conclusions into a forecast. No formal model is used, and no two experts are likely to consider the same information in the same way.

Expert judgment is a forecasting method that is commonly recommended when conditions in the past are not likely to hold in the future. Even though no formal quantitative model is used, expert judgment has provided good forecasts in many situations.

Scenario Writing

The qualitative procedure known as **scenario writing** consists of developing a conceptual scenario of the future based on a well-defined set of assumptions. Different sets of assumptions lead to different scenarios. The job of the decision maker is to decide how likely each scenario is and then to make decisions accordingly.

Intuitive Approaches

Subjective or *intuitive qualitative* approaches are based on the ability of the human mind to process a variety of information that, in most cases, is difficult to quantify. These techniques are often used in group work, wherein a committee or panel seeks to develop new ideas or solve complex problems through a series of "brainstorming sessions." In such sessions, individuals are freed from the usual group restrictions of peer pressure and criticism because they can present any idea or opinion without regard to its relevancy and, even more important, without fear of criticism.

SUMMARY

This chapter provided an introduction to the basic methods of time series analysis and forecasting. First, we showed that to explain the behavior of a time series, it is often helpful to think of the time series as consisting of four separate components: trend, cyclical, seasonal, and irregular. By isolating these components and measuring their apparent effect, one can forecast future values of the time series.

We discussed how smoothing methods can be used to forecast a time series that exhibits no significant trend, seasonal, or cyclical effect. The moving averages approach consists of computing an average of past data values and then using that average as the forecast for the next period. In the exponential smoothing method, a weighted average of past time series values is used to compute a forecast.

For time series that have only a long-term trend, we showed how regression analysis could be used to make trend projections. For time series in which both trend and seasonal influences are significant, we showed how to isolate the effects of the two factors and prepare better forecasts. Finally, regression analysis was described as a procedure for developing causal forecasting models. A causal forecasting model is one that relates the time series value (dependent variable) to other independent variables that are believed to explain (cause) the time series behavior.

Qualitative forecasting methods were discussed as approaches that could be used when little or no historical data are available. These methods are also considered most appropriate when the past pattern of the time series is not expected to continue into the future.

GLOSSARY

Time series   A set of observations measured at successive points in time or over successive periods of time.

Forecast   A projection or prediction of future values of a time series.

Trend   The long-run shift or movement in the time series observable over several periods of time.

Cyclical component   The component of the time series model that results in periodic above-trend and below-trend behavior of the time series lasting more than one year.

Seasonal component   The component of the time series model that shows a periodic pattern over one year or less.

Irregular component   The component of the time series model that reflects the random variation of the time series values beyond what can be explained by the trend, cyclical, and seasonal components.

Moving averages   A method of forecasting or smoothing a time series by averaging each successive group of data points.

Mean squared error (MSE)   An approach to measuring the accuracy of a forecasting model. This measure is the average of the sum of the squared differences between the forecast values and the actual time series values.

Weighted moving averages   A method of forecasting or smoothing a time series by computing a weighted average of past data values. The sum of the weights must equal one.

Exponential smoothing   A forecasting technique that uses a weighted average of past time series values to arrive at smoothed time series values that can be used as forecasts.

Smoothing constant   A parameter of the exponential smoothing model that provides the weight given to the most recent time series value in the calculation of the forecast value.

Multiplicative time series model   A model whereby the separate components of the time series are multiplied together to identify the actual time series value. When the four components of trend, cyclical, seasonal, and irregular are assumed present, we obtain $Y_t = T_t \times C_t \times S_t \times I_t$. When the cyclical component is not modeled, we obtain $Y_t = T_t \times S_t \times I_t$.

Deseasonalized time series   A time series from which the effect of season has been removed by dividing each original time series observation by the corresponding seasonal index.

Causal forecasting methods   Forecasting methods that relate a time series to other variables that are believed to explain or cause its behavior.

Autoregressive model   A time series model whereby a regression relationship based on past time series values is used to predict the future time series values.

Delphi method   A qualitative forecasting method that obtains forecasts through group consensus.

Scenario writing   A qualitative forecasting method that consists of developing a conceptual scenario of the future based on a well-defined set of assumptions.

KEY FORMULAS

Moving Average

$$\text{Moving Average} = \frac{\Sigma(\text{most recent } n \text{ data values})}{n} \tag{18.1}$$

Exponential Smoothing Model

$$F_{t+1} = \alpha Y_t + (1 - \alpha)F_t \tag{18.2}$$

Equation for Linear Trend

$$T_t = b_0 + b_1 t \tag{18.5}$$

Multiplicative Time Series Model With Trend, Seasonal, and Irregular Components

$$Y_t = T_t \times S_t \times I_t \tag{18.9}$$

Multiplicative Time Series Model With Trend, Cyclical, Seasonal, and Irregular Components

$$Y_t = T_t \times C_t \times S_t \times I_t \tag{18.10}$$

SUPPLEMENTARY EXERCISES

27. Moving averages often are used to identify movements in stock prices. Weekly closing prices (in dollars per share) for Toys R Us for September 22, 1997, through December 8, 1997, follow (Prudential Securities, Inc.).

Week	Price ($)	Week	Price ($)
September 22	$34\frac{7}{8}$	November 3	$33\frac{5}{8}$
September 29	$35\frac{5}{8}$	November 10	$35\frac{1}{16}$
October 6	$34\frac{11}{16}$	November 17	$34\frac{1}{16}$
October 13	$33\frac{9}{16}$	November 24	$34\frac{1}{8}$
October 20	$32\frac{5}{8}$	December 1	$33\frac{1}{4}$
October 27	34	December 8	$32\frac{1}{16}$

a. Use a 3-month moving average to smooth the time series. Forecast the closing price for December 15, 1997.

b. Use a 3-month weighted moving average to smooth the time series. Use a weight of .4 for the most recent period, .4 for the next period back, and .2 for the third period back. Forecast the closing price for December 15, 1997.

c. Use exponential smoothing with a smoothing constant of $\alpha = .35$ to smooth the time series. Forecast the closing price for December 15, 1997.

d. Which of the three methods do you prefer? Why?

28. The Federal Election Commission maintains data showing the voting age population, the number of registered voters, and the turnout for federal elections. The following table shows the national voter turnout as a percentage of the voting age population from 1972 to 1996. (*The Wall Street Journal Almanac,* 1998)

Year	% Turnout	Year	% Turnout	Year	% Turnout
1972	55	1982	40	1990	37
1974	38	1984	53	1992	55
1976	54	1986	36	1994	39
1978	37	1988	50	1996	49
1980	53				

   a. Use exponential smoothing to forecast this time series. Consider smoothing constants of $\alpha = .1$ and .2. What value of the smoothing constant provides the best forecast?

   b. What is the forecast of the percentage turnout in 1998?

29. The percentage of individual investors' portfolios committed to stock depends on the state of the economy. As of April 1997, a typical portfolio consisted of cash (19%), stocks (30%), stock funds (37%), bonds (8%), and bond funds (6%) (*AAII Journal,* June 1997). The following table reports the percentage of stocks in a typical portfolio in nine quarters from 1995 to 1997.

Quarter	Stock %
1st—1995	29.8
2nd—1995	31.0
3rd—1995	29.9
4th—1995	30.1
1st—1996	32.2
2nd—1996	31.5
3rd—1996	32.0
4th—1996	31.9
1st—1997	30.0

   a. Use exponential smoothing to forecast this time series. Consider smoothing constants of $\alpha = .2, .3,$ and .4. What value of the smoothing constant provides the best forecast?

   b. What is the forecast of the percentage of assets committed to stocks for the second quarter of 1997?

30. A chain of grocery stores noted the weekly demand (in cases) reported in the following table for a particular brand of automatic dishwasher detergent. Use exponential smoothing with $\alpha = .2$ to develop a forecast for week 11.

Week	Demand
1	22
2	18
3	23
4	21
5	17
6	24
7	20
8	19
9	18
10	21

31. United Dairies, Inc., supplies milk to several independent grocers throughout Dade County, Florida. Managers at United Dairies want to develop a forecast of the number of half-gallons of milk sold per week. Sales data for the past 12 weeks follow.

Week	Sales	Week	Sales
1	2750	7	3300
2	3100	8	3100
3	3250	9	2950
4	2800	10	3000
5	2900	11	3200
6	3050	12	3150

Using exponential smoothing with $\alpha = .4$, develop a forecast of demand for the 13th week.

32. The Garden Avenue Seven sells tapes of its musical performances. The following table reports sales (in units) for the past 18 months. The group's manager wants an accurate method for forecasting future sales.

Month	Sales	Month	Sales	Month	Sales
1	293	7	381	13	549
2	283	8	431	14	544
3	322	9	424	15	601
4	355	10	433	16	587
5	346	11	470	17	644
6	379	12	481	18	660

a. Use exponential smoothing with $\alpha = .3, .4$, and $.5$. Which value of α provides the best forecasts?

b. Use trend projection to provide a forecast. What is the value of MSE?

c. Which method of forecasting would you recommend to the manager? Why?

33. The Mayfair Department Store in Davenport, Iowa, is trying to determine the amount of sales lost while it was shut down during July and August because of damage caused by the Mississippi River flood. Sales data for January through June follow.

Month	Sales ($1000s)	Month	Sales ($1000s)
January	185.72	April	210.36
February	167.84	May	255.57
March	205.11	June	261.19

a. Use exponential smoothing, with $\alpha = .4$, to develop a forecast for July and August. (Hint: Use the forecast for July as the actual sales in July in developing the August forecast.) Comment on the use of exponential smoothing for forecasts more than one period into the future.

b. Use trend projection to forecast sales for July and August.

c. Mayfair's insurance company has proposed a settlement based on lost sales of $240,000 in July and August. Is this amount fair? If not, what amount would you recommend as a counteroffer?

34. Canton Supplies, Inc., is a service firm that employs approximately 100 individuals. Managers of Canton Supplies are concerned about meeting monthly cash obligations and want to develop a forecast of monthly cash requirements. Because of a recent change in operating policy, only the past seven months of data are considered to be relevant. With the following historical data, use trend projection to develop a forecast of cash requirements for each of the next two months.

Month	1	2	3	4	5	6	7
Cash Required ($1000s)	205	212	218	224	230	240	246

35. The following table reports the number of cable television subscribers (1000s) by year from 1980 to 1994 (*Television & Cable Factbook*, 1994).

Year	Number	Year	Number
1980	16,000	1988	44,000
1981	18,300	1989	47,500
1982	21,000	1990	50,000
1983	25,000	1991	51,000
1984	30,000	1992	53,000
1985	32,000	1993	55,000
1986	37,500	1994	57,000
1987	41,100		

   a. Develop a linear trend equation for this time series. Comment on what the equation tells about the number of cable subscribers for the 15-year period.

   b. Provide forecasts of the number of cable subscribers for 1995 and 1996.

36. The Costello Music Company has been in business for 5 years. During that time, sales of electric organs have increased from 12 units in the first year to 76 units in the most recent year. Fred Costello, the firm's owner, wants to develop a forecast of organ sales for the coming year. The historical data follow.

Year	1	2	3	4	5
Sales	12	28	34	50	76

   a. Show a graph of this time series. Does a linear trend appear to be present?

   b. Develop the equation for the linear trend component for the time series. What is the average increase in sales that the firm has been realizing per year?

37. Hudson Marine has been an authorized dealer for C&D marine radios for the past 7 years. The following table reports the number of radios sold each year.

Year	1	2	3	4	5	6	7
Number Sold	35	50	75	90	105	110	130

   a. Show a graph of this time series. Does a linear trend appear to be present?

   b. Develop the equation for the linear trend component of the time series.

   c. Use the linear trend developed in part (b) to develop a forecast for annual sales in year 8.

38. The Motion Picture Association of America (MPAA) collects data that show the cost of making movies for MPAA members, including Disney, Paramount, Universal, Warner Brothers, MGM, Fox, Sony, and Turner. The following data show the average production costs in thousands of dollars from 1985 to 1996 for MPAA members (*The Wall Street Journal Almanac,* 1998).

Year	Average Production Costs ($1000s)	Year	Average Production Costs ($1000s)
1985	16,779	1991	26,136
1986	17,455	1992	28,858
1987	20,051	1993	29,910
1988	18,061	1994	34,288
1989	23,454	1995	36,370
1990	26,783	1996	39,836

   a. Plot the time series and comment on the appropriateness of a linear trend.

   b. Develop the equation for the linear trend component of this time series.

   c. What is the average increase in production costs that MPAA members have been experiencing per year?

   d. Use the trend equation to forecast the average production costs in 1997 and 1998.

39. The following table gives the number of cellular telephone subscribers (1000s) by year from 1986 to 1994 (*State of the Cellular Industry,* 1994).

Year	Number	Year	Number
1986	682	1990	5,283
1987	1,231	1991	7,557
1988	2,069	1992	11,033
1989	3,509	1993	16,009

Plot the time series and comment on the appropriateness of a linear trend. What type of functional form do you believe would be most appropriate for the trend pattern of this time series?

40. Refer to the Hudson Marine problem in Exercise 37. Suppose the quarterly sales values for the 7 years of historical data are as follow.

Year	Quarter 1	Quarter 2	Quarter 3	Quarter 4	Total Yearly Sales
1	6	15	10	4	35
2	10	18	15	7	50
3	14	26	23	12	75
4	19	28	25	18	90
5	22	34	28	21	105
6	24	36	30	20	110
7	28	40	35	27	130

   a. Show the 4-quarter moving average values for this time series. Plot both the original time series and the moving average series on the same graph.
   b. Compute the seasonal indexes for the four quarters.
   c. When does Hudson Marine experience the largest seasonal effect? Does this seem reasonable? Explain.

41. Consider the Costello Music Company problem in Exercise 36. The quarterly sales data follow.

Year	Quarter 1	Quarter 2	Quarter 3	Quarter 4	Total Yearly Sales
1	4	2	1	5	12
2	6	4	4	14	28
3	10	3	5	16	34
4	12	9	7	22	50
5	18	10	13	35	76

   a. Compute the seasonal indexes for the four quarters.
   b. When does Costello Music experience the largest seasonal effect? Does this appear reasonable? Explain.

42. Refer to the Hudson Marine data in Exercise 40.
   a. Deseasonalize the data and use the deseasonalized time series to identify the trend.
   b. Use the results of part (a) to develop a quarterly forecast for next year based on trend.
   c. Use the seasonal indexes developed in Exercise 40 to adjust the forecasts developed in part (b) to account for the effect of season.

43. Consider the Costello Music Company time series in Exercise 41.
   a. Deseasonalize the data and use the deseasonalized time series to identify the trend.
   b. Use the results of part (a) to develop a quarterly forecast for next year based on trend.
   c. Use the seasonal indexes developed in Exercise 41 to adjust the forecasts developed in part (b) to account for the effect of season.

Case Problem 1   FORECASTING FOOD AND BEVERAGE SALES

The Vintage Restaurant is on Captiva Island, a resort community near Fort Myers, Florida. The restaurant, which is owned and operated by Karen Payne, has just completed its third year of operation. During that time, Karen has sought to establish a reputation for the restaurant as a high-quality dining establishment that specializes in fresh seafood. The efforts by Karen and her staff have proven successful, and her restaurant has become one of the best and fastest-growing restaurants on the island.

Karen has concluded that to plan for the growth of the restaurant in the future, she needs to develop a system that will enable her to forecast food and beverage sales by month for up to one year in advance. Karen has the following data ($1000s) on total food and beverage sales for the three years of operation.

Month	First Year	Second Year	Third Year
January	242	263	282
February	235	238	255
March	232	247	265
April	178	193	205
May	184	193	210
June	140	149	160
July	145	157	166
August	152	161	174
September	110	122	126
October	130	130	148
November	152	167	173
December	206	230	235

Managerial Report

Perform an analysis of the sales data for the Vintage Restaurant. Prepare a report for Karen that summarizes your findings, forecasts, and recommendations. Include:

1. A graph of the time series.
2. An analysis of the seasonality of the data. Indicate the seasonal indexes for each month, and comment on the high and low seasonal sales months. Do the seasonal indexes make intuitive sense? Discuss.
3. A forecast of sales for January through December of the fourth year.
4. Recommendations as to when the system that you have developed should be updated to account for new sales data.
5. Any detailed calculations of your analysis in the appendix of your report.

Assume that January sales for the fourth year turn out to be $295,000. What was your forecast error? If this is a large error, Karen may be puzzled about the difference between your forecast and the actual sales value. What can you do to resolve her uncertainty in the forecasting procedure?

Case Problem 2   FORECASTING LOST SALES

The Carlson Department Store suffered heavy damage when a hurricane struck on August 31, 2000. The store was closed for four months (September 2000 through December 2000), and Carlson is now involved in a dispute with its insurance company about the amount of lost sales during the time the store was closed. Two key issues must be resolved: (1) the amount of sales Carlson would have made if the hurricane had not struck and (2) whether Carlson is entitled to any compensation for excess sales due to increased business activity after the storm. More than $8 billion in federal disaster relief and insurance money came into the county, resulting in increased sales at department stores and numerous other businesses.

Table 18.13 gives Carlson's sales data for the 48 months preceding the storm. Table 18.14 reports total sales for the 48 months preceding the storm for all department stores in the county, as well as the total sales in the county for the four months the Carlson Department Store was closed. Carlson's managers have asked you to analyze these data and develop estimates of the lost sales at the Carlson Department Store for the months of September through December 2000. They also have asked you to determine whether a case can be

TABLE 18.13   SALES FOR CARLSON DEPARTMENT STORE, SEPTEMBER 1996 THROUGH AUGUST 2000 ($ MILLIONS)

Month	1996	1997	1998	1999	2000
January		1.45	2.31	2.31	2.56
February		1.80	1.89	1.99	2.28
March		2.03	2.02	2.42	2.69
April		1.99	2.23	2.45	2.48
May		2.32	2.39	2.57	2.73
June		2.20	2.14	2.42	2.37
July		2.13	2.27	2.40	2.31
August		2.43	2.21	2.50	2.23
September	1.71	1.90	1.89	2.09	
October	1.90	2.13	2.29	2.54	
November	2.74	2.56	2.83	2.97	
December	4.20	4.16	4.04	4.35	

made for excess storm-related sales during the same period. If such a case can be made, Carlson is entitled to compensation for excess sales it would have earned in addition to ordinary sales.

Managerial Report

Prepare a report for the managers of the Carlson Department Store that summarizes your findings, forecasts, and recommendations. Include:

1. An estimate of sales had there been no hurricane.
2. An estimate of countywide department store sales had there been no hurricane.
3. An estimate of lost sales for the Carlson Department Store for September through December 2000.

In addition, use the countywide actual department stores sales for September through December 2000 and the estimate in part (2) to make a case for or against excess storm-related sales.

TABLE 18.14   DEPARTMENT STORE SALES FOR THE COUNTY, SEPTEMBER 1996 THROUGH DECEMBER 2000 ($ MILLIONS)

Month	1996	1997	1998	1999	2000
January		46.8	46.8	43.8	48.0
February		48.0	48.6	45.6	51.6
March		60.0	59.4	57.6	57.6
April		57.6	58.2	53.4	58.2
May		61.8	60.6	56.4	60.0
June		58.2	55.2	52.8	57.0
July		56.4	51.0	54.0	57.6
August		63.0	58.8	60.6	61.8
September	55.8	57.6	49.8	47.4	69.0
October	56.4	53.4	54.6	54.6	75.0
November	71.4	71.4	65.4	67.8	85.2
December	117.6	114.0	102.0	100.2	121.8

Appendix 18.1   **FORECASTING WITH MINITAB**

In this appendix we show how Minitab can be used to develop forecasts using three forecasting methods: moving averages, exponential smoothing, and trend projection.

Moving Averages

Gasoline

To show how Minitab can be used to develop forecasts using the moving averages method, we will develop a forecast for the gasoline sales time series in Table 18.1 and Figure 18.5. The sales data for the 12 weeks have been entered into column 2 of the worksheet. The following steps can be used to produce a 3-week moving average forecast for week 13.

Step 1. Select the **Stat** pull-down menu
Step 2. Choose **Time Series**
Step 3. Choose **Moving Average**
Step 4. When the Moving Average dialog box appears:
Enter C2 in the **Variable box**
Enter 3 in the **MA length** box
Select **Generate forecasts**
Enter 1 in the **Number of forecasts** box
Enter 12 in the **Starting from origin** box
Click **OK**

The 3-week moving average forecast for week 13 is shown in the session window. The mean square error of 10.22 is labeled as MSD in the Minitab output. Many other output options are available, including a summary table similar to Table 18.2 and graphical output similar to Figure 18.6.

Exponential Smoothing

Gasoline

To show how Minitab can be used to develop an exponential smoothing forecast, we will again develop a forecast of sales in week 13 for the gasoline sales time series in Table 18.1 and Figure 18.5. The sales data for the 12 weeks have been entered into column 2 of the worksheet. The following steps can be used to produce a forecast for week 13 using a smoothing constant of $\alpha = .2$.

Step 1. Select the **Stat** pull-down menu
Step 2. Choose **Time Series**
Step 3. Choose **Single Exp Smoothing**
Step 4. When the Single Exponential Smoothing dialog box appears:
Enter C2 in the **Variable** box
Select the **Use** option for the Weight to Use in Smoothing
Enter 0.2 in the **Use** box
Select **Generate forecasts**
Enter 1 in the **Number of forecasts** box
Enter 12 in the **Starting from origin** box
Select **Options**
Step 5. When the Single Exponential Smoothing—Options dialog box appears:
Enter 1 in the **Use average of first** box
Click **OK**
Step 6. When the Single Exponential Smoothing dialog box appears:
Click **OK**

The exponential smoothing forecast for week 13 is shown in the session window. The mean square error is labeled as MSD in the Minitab output.* Many other output options are available, including a summary table similar to Table 18.3 and graphical output similar to Figure 18.7.

Trend Projection

CD file

Bicycle

To show how Minitab can be used for trend projection, we develop a forecast for the bicycle sales time series in Table 18.6 and Figure 18.8. The year numbers have been entered into column C1 and the sales data have been entered into column C2 of the worksheet. The following steps can be used to produce a forecast for week 13 using trend projection.

Step 1. Select the **Stat** pull-down menu
Step 2. Choose **Time Series**
Step 3. Choose **Trend Analysis**
Step 4. When the Trend Analysis dialog box appears:
  Enter C2 in the **Variable** box
  Choose **Linear** for the Model Type
  Select **Generate forecasts**
  Enter 1 in the **Number of forecasts** box
  Enter 10 in the **Starting from origin** box
  Click **OK**

The equation for linear trend and the forecast for the next period are shown in the session window.

Appendix 18.2   FORECASTING WITH EXCEL

In this appendix we show how Excel can be used to develop forecasts using three forecasting methods: moving averages, exponential smoothing, and trend projection.

Moving Averages

To show how Excel can be used to develop forecasts using the moving averages method, we will develop a forecast for the gasoline sales time series in Table 18.1 and Figure 18.5. The sales data for the 12 weeks have been entered into worksheet rows 2 through 13 of column B. The following steps can be used to produce a 3-week moving average.

Step 1. Select the **Tools** pull-down menu
Step 2. Choose **Data Analysis**
Step 3. Choose **Moving Average** from the list of Analysis Tools
  Click **OK**
Step 4. When the Moving Average dialog box appears:
  Enter B2:B13 in the **Input Range** box
  Enter 3 in the **Interval** box
  Enter C2 in the **Output Range** box
  Click **OK**

The 3-week moving average forecasts will appear in column B of the worksheet. Forecasts for periods of other length can be computed easily by entering a different value in the **Interval** box.

---

*The value of MSD computed by Minitab is not the same as the value of MSE that appears in Table 18.4. Minitab uses a forecast of 17 for week 1 and computes MSD using all 12 time periods of data. In Section 18.2 we compute MSE using only the data for weeks 2 through 12, because we had no past values with which to make a forecast for period 1.

Exponential Smoothing

Gasoline

To show how Excel can be used for exponential smoothing, we again develop a forecast for the gasoline sales time series in Table 18.1 and Figure 18.5. The sales data for the 12 weeks have been entered into worksheet rows 2 through 13 of column B. The following steps can be used to produce a forecast using a smoothing constant of $\alpha = .2$.

Step 1. Select the **Tools** pull-down menu
Step 2. Choose **Data Analysis**
Step 3. Choose **Exponential Smoothing** from the list of Analysis Tools
Click **OK**
Step 4. When the Exponential Smoothing dialog box appears:
Enter B2:B13 in the **Input Range** box
Enter .8 in the **Damping factor** box
Enter C2 in the **Output Range** box
Click **OK**

The exponential smoothing forecasts will appear in column B of the worksheet. Note that the value we entered in the Damping factor box is $1 - \alpha$; forecasts for other smoothing constants can be computed easily by entering a different value for $1 - \alpha$ in the Damping factor box.

Trend Projection

Bicycle

To show how Excel can be used for trend projection, we develop a forecast for the bicycle sales time series in Table 18.6 and Figure 18.8. The data, with appropriate labels in row 1, have been entered into worksheet rows 1 through 11 of columns A and B. The following steps can be used to produce a forecast for year 11 by trend projection.

Step 1. Select an empty cell in the worksheet
Step 2. Select the **Insert** pull-down menu
Step 3. Choose **Function**
Step 4. When the Paste Function dialog box appears:
Choose **Statistical** in the Function Category box
Choose **Forecast** in the Function Name box
Click **OK**
Step 5. When the Forecast dialog box appears:
Enter 11 in the **x** box
Enter B2:B11 in the **Known y's** box
Enter A2:A11 in the **Known x's** box
Click **OK**

The forecast for year 11, in this case 32.5, will appear in the cell selected in Step 1.